GEOMETRY AND MEANING

CSLI LECTURE NOTES NO. 172

GEOMETRY AND MEANING

Dominic Widdows

Foreword by Pentti Kanerva

CSLI PUBLICATIONS
Center for the Study of
Language and Information
Stanford, California

Copyright © 2004
CSLI Publications
Center for the Study of Language and Information
Leland Stanford Junior University
Printed in the United States
08 07 06   2 3 4 5

Library of Congress Cataloging-in-Publication Data

Widdows, Dominic, 1974–
Geometry and meaning / by Dominic Widdows.

p. cm. – (CSLI lecture notes ; no. 172)

Includes bibliographical references and index.

ISBN 1-57586-448-7 (pbk. : alk. paper)
ISBN 1-57586-447-9 (hardback : alk. paper)

1. Geometry. 2. Graph theory. 3. Group schemes (Mathematics)
4. Mathematical linguistics. I. Title. II. Series.

QA564.W53 2004
516–dc22   2004008661
CIP

∞ The acid-free paper used in this book meets the minimum requirements of the American National Standard for Information Sciences—Permanence of Paper for Printed Library Materials, ANSI Z39.48-1984.

CSLI was founded in 1983 by researchers from Stanford University, SRI International, and Xerox PARC to further the research and development of integrated theories of language, information, and computation. CSLI headquarters and CSLI Publications are located on the campus of Stanford University.

CSLI Publications reports new developments in the study of language, information, and computation. Please visit our web site at
http://cslipublications.stanford.edu/
for comments on this and other titles, as well as for changes and corrections by the author and publisher.

Dedication

This book, and a grove of trees near Bromyard in Herefordshire, is dedicated with loving memory to David Henry, friend to my family from before I was born, who, dying of cancer, introduced a young mathematician to the concept of natural language processing.

Contents

Foreword   **ix**

Introduction   **xvi**

1   Geometry, Numbers and Sets   **1**

2   The Graph Model: Networks of Concepts   **41**

3   The Vertical Direction: Concept Hierarchies   **68**

4   Measuring Similarity and Distance   **98**

5   Word-Vectors and Search Engines   **132**

6   Exploring Vector Spaces in One and More Languages   **167**

7   Logic with Vectors: Ambiguous Words and Quantum States   **200**

8   Concept Lattices: Binding Everything Loosely Together   **246**

Conclusions and Curtain Call   **289**

References   **304**

Index   **314**

Foreword
by Pentti Kanerva

We think of some sciences, by their very nature, as being more mathematical than others. We even refer to physics and other heavily mathematical sciences as "hard sciences," as if to imply that social sciences and humanities are somehow soft or easy. Another way to see this picture, however, is that mathematics itself is hard and that it would be applied first to sciences that themselves are relatively easy. Let us stay with this idea for a moment.

Mathematics is hard, in the sense that it is not in our nature. Like reading and writing, or playing a musical instrument, it takes a great deal of practice to become good at it. Mathematics is also an ever-growing field that builds on its earlier discoveries. Thus it seems reasonable that we would discover the simpler parts of math first and that they would be sufficient for describing phenomena that are relatively simple. However, the thoroughness with which mathematics has imbued the "hard" sciences testifies of its power to aid our thinking once we have found the appropriate mathematics.

It matters greatly that the mathematics be appropriate, and there are conspicuous counterexamples. Early math dealt with numbers, ratios, and geometric shapes, and these were sufficient for surveying the land and describing musical sounds pleasing to the ear. Attempts to explain the wandering planets or the basic elements of nature by the five perfect solids, however, have failed the test of time and earned the name pseudoscience. The math that we now use to explain the motion of the planets was only discovered two millennia later.

Once a field finds its math, they begin to feed on each other and coevolve. Old scientific puzzles get satisfactory explanations, new puzzles inspire further developments in math, and the math points to new questions in the science—to what else might be true and worth investigating. That is how physics, for example, has become so thoroughly mathematical.

Can this happen with anything as rich, varied, and complex as human thought and language? My guess is that it will once the appropriate math has been found, and the key is that it be appropriate. This may sound circular and therefore needs clarification.

The common notion that mathematics is about calculating and proving theorems is far too limited. Mathematics is the study of patterns: of things that repeat and recur in different contexts, of structures and relations that are common to different fields and situations. Mathematics seeks regularities in the behavior of natural and artificial systems and so it is the abstract study of things that are understandable to us—assuming that we are finite physical beings and thereby limited in our understanding.

Mathematics is a tool for coping with complexity. The world is infinitely complex from the human perspective, but the more regularity we find in it the better sense it makes to us. Finding the regularities is the key, and dressing them into theorems and proofs is mostly a style of communication among mathematicians. This style is relatively safe and effective, although it belies the process by which mathematicians arrive at their results. Such a style can be compared to rhyming in poetry: how rhyme helps us remember the verse. What matters is the idea behind the theorem and its proof, as is the image or feeling behind the verse.

The mathematics for describing the motions of planets allows us to calculate their positions for centuries to come, whereas no amount of calculation will let us predict exactly what a person will say or do tomorrow. This is commonly interpreted in one of two ways, either that mathematics does not apply after all, or that the proper math here would be probability because we can at least guess what a person might say or do.

The first interpretation reflects the overly limited view of mathematics noted above and can be dismissed on those grounds. The second ascribes too narrow a role to mathematics and thus misses the mark. It overemphasizes prediction—if we cannot predict exactly then let us predict probabilistically—and assumes that the needed math is already fully developed and need only be applied. In language, for example, central aspects such as structure and meaning require their own kind of math in which probability plays only a side role. Much of that math may not even exist yet but will be developed concurrently with our increasing understanding of structure and meaning.

This brings us to the substance of Dominic's book: the exploration of mathematics that would be appropriate for describing concepts and meaning. The branch of mathematics that has traditionally been associated with meaning is logic. Logic is also discussed here, yet the main emphasis is on identifying mathematical spaces that would lend to modeling of meaning.

High-dimensional vectors make up one such space—or, rather, a family of spaces—and the distance between two vectors is a basic notion of a vector space. Distance, or closeness, is a geometric notion and so we enter the realm of geometry. When words are represented by vectors, the closeness of vectors can represent the similarity of their meanings. In mathematics, the elements that make up a space are called points of that space. Here the points are vectors, but a distance between points is a valid idea in many mathematical spaces and thus can model similarity of meaning more generally.

Probability becomes a part of the vectors when they are based on word frequencies in large bodies of text. The essence of this book, however, is not probability itself but the mathematical structures that the probabilities reflect, such as graphs and lattices. By matching more closely the mind's underlying mechanisms, these structures can provide deeper insight into language and meaning than would the probabilities by themselves. With regard to logic, the vectors allow traditional logics to be generalized to "quantum" logics that correspond more closely to how words are actually used.

The human sciences are young from mathematics' perspective, and extremely challenging problems abound. For example, how do words and language expressions acquire meaning? They become meaningful to us in various ways: from contexts in which they appear, from examples of their usage, and by being defined in terms of other words and expressions, as in a dictionary. Meanings are also transferred readily to new domains by analogy and metaphor. But what representations and underlying mechanisms—what physical processes—would have these properties? What mathematical structures would show similar behavior and allow it to be analyzed in depth?

Everyday language is transparent to us but when we try to program computers to use it, it proves to be endlessly ambiguous. Obviously, how we represent language in computers—that is, our mathematical model of language—is grossly inadequate. A more faithful model would have great scientific and practical value.

Our ability to learn language is one of nature's marvels. It gets credited to innateness or instinct, which no doubt is correct, but we should not stop there. We should also ask, what are the underlying mechanisms, and what kind of a model would adequately explain them? The model would be a mathematical abstraction of how the brain's circuits work, but the mathematics to describe the model may not even exist yet.

Complex learning in general is poorly understood in terms of underlying mechanisms. Much of it takes place in social settings in which we observe and absorb into ourselves the behavior of others, and this kind of learning is important in the animal world at large. A mathematical model that ties learning to underlying mechanisms would deepen our understanding of ourselves as social beings. It would also take us a long way toward the building of artificial systems that learn and adapt in ways that humans and animals do.

I may seem to be overselling mathematical modeling, but I am emphasizing it for a reason. We humans are very limited in our ability to imagine and see things in a new light. Even when our concepts and explanations are badly wanting, we tend to hold onto them. Anything that can help us surmount our limitations should therefore be more than welcome. We fare much better when we bootstrap by using analogy—that is, relate the new to something we already know, or model the unfamiliar with the old and familiar.

Here is where mathematics can help. It is a rich source of models, capable of spurring new branches of math and new models to capture whatever regularity interests us, and so it has the potential of serving any field of study. A crucial part is still left for human creativity, namely, seeing some behavior and imagining an underlying system that could produce it. The deeper our knowledge of the behavior and the richer our repertoire of models, the more likely we are to make a meaningful connection. In the interplay that follows, the behavior suggests extensions of the math and the model, and the model predicts new facets of the behavior. When the right conditions for discovery are provided, a stroke of genius is not merely a stroke of luck any more.

This book is a mathematician's invitation for researchers in the human sciences to look into new areas of math, and it gives an illustrated tour with many examples from language. It is not a book of techniques—there are other sources for that—but of the ideas behind the techniques that would be helpful for understanding existing techniques and for discovering new ones. This is useful to us all, yet to young people in particular I recommend an even broader study of mathematics so that you can make it serve you on your terms and not be limited by what others have done already. To use a military metaphor, you want this arsenal at your disposal. The book can also help mathematicians to identify challenging and worthy problems to which to apply their skills.

The writing is congenial, which is rare among texts in mathematics. It conveys the author's appreciation of, and his eagerness to share with others, the beauty in the subject and the humanity in the people who have brought it to us over the millennia. By going to original writings he has discovered how the story has changed in the telling by those who did not fully understand it. This has allowed him to repair the reputation, as a mathematician, of even someone like Aristotle.

I will close with a bold challenge: mathematics is too important—it is too useful in solving fundamental problems in fields other than math—to be left solely to mathematicians. Social scientists, linguists, philosophers, mathematicians, all have much to contribute and gain.

Some major historical and mathematical trends which form a background to this book.

— 3000 BC

Great Pyramids built in Egypt

Egyptian calendar introduced (possibly earlier)
Egyptian geometry (the rope–sretchers)

Babloynian number system, positional notation

Megalithic monuments built in Europe

— 2000 BC

Turbulent 'Dark Age' as iron supersedes bronze

— 1000 BC

Greek city–states emerge, borrow alphabet from near East

Conquests of Alexander spread Greek language and scholarship

} Greek geometry, logic and biology
(See detail overleaf)

Progress in Greek trigonometry and measurement

Roman Empire, finally splits into Byzantine (Greek speaking) East and Latin West, which collapses

— 1 BC / 1 AD

Decline of mathematics in the West

Dark Age in Western Europe

Islamic culture spreads from Spain to Central Asia; considerable contact with India and China

Progress in algebra and trigonometry in India and the Islamic world

Al–Kwarizmi publishes 'Al–Jabr' (algebra)

— 1000 AD

Mongol invasions decimate Islamic culture

Renaissance of learning in Europe

European technology and cultural influence spread by colonization

Hindu numerals (mistakenly called 'Arabic') make gradual progress in Europe

} Western progress in geometry, logic algebra, biology and computing
(See detail overleaf)

— 2000 AD

Scientific progress in two particularly important periods.

Ancient Greek scientific progress.

- 600 BC

  First recorded geometric proofs
  due to Thales of Miletus

  Pythagoras founds school in Italy

  Crisis of incommensurables in number theory
- 500 BC

  Atomism introduced by Leucippus
  Empedocles develops theory of four elements
  Zeno presents paradoxes of motion
- 400 BC

  Eudoxus develops method of proportions
  Plato develops geometric theory
  of creation and elements
  Journeys and works of Aristotle

- 300 BC
  **Euclid of Alexandria writes the
  Elements of Geometry**

  Aristarchus measures distance to sun and moon

  Eratosthenes measures circumference of earth
  Appolonius uses a form of coordinates
- 200 BC

Modern scientific progress.

- 1500

  Copernicus describes solar system

  Stevin introduces decimal comma
- 1600
  Kepler's laws of planetary motion

  Descartes' Analytic Geometry
  Pascal introduces probability

  Newton's laws of gravitation and motion
- 1700

  Euler introduces graph theory
  Samuel Johnson publishes Dictionary

- 1800

  Hamilton, Grassmann introduce vectors
  Boole develops mathematical logic
  Roget publishes thesaurus
  Darwin describes evolution of species

  Axioms of set theory and number theory
- 1900

  Relativity and quantum theory
  revolutionize physics
  Birkhoff and von Neumann introduce quantum logic
  Computers invented
  Shannon discovers information theory
  Lattice theory matures
  Invention of internet
- 2000

Introduction

The year Plato died, his most versatile pupil left the prestigious Academy in cosmopolitan Athens, and embarked upon a 12 year scientific field-trip around the Aegean sea, studying a great variety of species and classifying his discoveries as best he could into a regular, coherent structure. His name was Aristotle. Two generations later, a lecturer at the new University in Alexandria wrote a textbook called the *Elements of Geometry*, presenting the most important mathematical results known to date in a unified form, deduced systematically from a small collection of basic assumptions. His name was Euclid.

These towering figures, who came to personify the studies of meaning and of geometry in Western tradition, would have been fascinated by the spectacular applications of their methods in the internet age, with its flowering of geometric representations for information of all kinds. Many of these structures are very apparent in our daily lives, from the maps of public transportation networks to the graphical interfaces of computer systems and browsers which enable users to click on files and documents like points on a map, linking them to vast networks of information. These perceptible examples are the tip of an iceberg. Behind the scenes, many search engines represent queries and documents as points in geometric spaces and trawl their way through the internet using the patterns of hyperlinks to measure the reliability of different information sources. Knowledge bases represent linguistic concepts in hierarchical tree-structures, which enable logical inferences to combine different premises and deduce valid and relevant conclusions.

Many of the geometric techniques applied in recent years to

information technology are centuries old, and have been developed through contact with mechanics, electronics, acoustics, biology, relativity and quantum mechanics. From ancient and mystical correspondences between shapes, numbers, musical notes and physical elements, to the revolutions of relativity and quantum theory in recent times, this book tells the story of the geometric spaces which provide the mathematical stage on which the scientific action takes place, and how many of their principles are derived straight from the ancient works of Euclid and Aristotle.

Originating from many branches of science, all of these spaces are used to represent words and their meanings, and the book presents many varied examples of these uses. Most of these are taken from models built automatically from free text, using software and models developed by the Infomap (*Info*rmation *Map*ping) project, a team at Stanford University's Center for the Study of Language and Information which I had the privilege of leading for three years, from 2001 to 2004. Many of these models can be used interactively over the internet, and I strongly recommend readers with internet access to go to our website[1] and explore a few of their favourite words and concepts for themselves — seeing is believing.

This story combines old and new. Ongoing progress in information management, vital to the needs of business, research, governmental and international organizations, is described right alongside the geometric spaces underlying the whole process, which have ancient roots. Rather than presenting the formal ideas as disembodied mathematical definitions and equations (which can be very dismaying to most human readers), I have tried to describe the circumstances and motivations which led to the original development of the geometric models we use for gathering information from text today. It is hoped that in so doing, readers who are new to natural language processing but have a grounding in other fields such as biology, music, physics, artificial intelligence, logic and cognitive science, will find some of the material strikingly familiar and may wonder if our methods have simply been stolen wholesale from other disciplines. In many cases this is just what has happened — mathematics has always seen the virtue in recycling.

As well as being a monograph of recent research and a historical

---

[1] http://infomap.stanford.edu/webdemo

perspective thereon, there is an important theme to this book. The coordinate geometry of René Descartes (1596–1650) revolutionized mathematics, bringing geometry and algebra together into a new combined structure. This catalyzed the development of scientific advances as varied and important as mechanics, gravitation, fluid dynamics, and electromagnetism, and provided the background within which Boole, Hamilton and Grassmann developed their own conceptual models in the 1840s and 1850s, which are used in search engines today. By the 1930s, mathematicians and quantum theorists began to recognize that geometry and *logic* could be similarly combined, within the common framework of lattice theory.

Last year, I programmed some of these logical operations into a geometric search engine and published the results of these experiments, which were both promising and thought provoking. The behaviour of these abstract models leads to important conceptual questions, some of which also lie at the heart of one of the greatest unsolved problems in physics, the discrepancy between classical physics and the quantum theory. As the history of science unfolds, the eventual union of geometry and logic could be as great a catalyst as the union of geometry and algebra wrought by Descartes over 350 years ago, describing the states of physical systems, states of information, and states of mind, providing frameworks for technological advances yet unthought of.

Who should read this book?

Geometry is a branch of mathematics, and you may firmly believe that there's no way you could read a book about mathematics and both understand the material and enjoy the experience. This is wrong, and the way mathematics is usually presented is very much to blame. If you have an eagerness for science and an inquisitive mind, you will understand and enjoy as much of this book as you care to read, and I hope you'll enjoy the pictures even in the sections you skip over.

As well as those with general scientific interests, this book may be useful to active researchers in many fields as a handbook of geometric models with worked examples (and some free software). Most of the examples are from natural language processing, but many of the models have been used in a variety other fields such as linguistics, computer science, artificial intelligence, sociology, psychology and cognitive science, where accurately modelling and exploring empirical data is an increasingly important activity. On the other hand, mathematicians who have been taught much of the theory behind geometric spaces might be delighted to find out about how these spaces can be used so elegantly and efficiently for solving problems in information engineering, and ideally they may even be attracted to a whole new career where their skills are relevant in ways they had not previously realized.

At its most specific, the book can be used to teach a graduate course in language and informatics: some parts have been successfully used already in a new course on Computational Word Learning at Stanford, which attracted students from many academic departments and researchers from local industry. Exercises (in the form of mathematical problems, programming challenges and written essays) can be provided for students at many different levels, and are available from the author on request.

However, my dearest (and for a mathematical book, most ambitious) desire has been that *Geometry and Meaning* should be enjoyable. Many people with excellent backgrounds in the sciences or humanities still feel that computers are a modern mystery which has left them behind. Many with a wonderful grasp of natural geometry were so firmly put off mathematics in high school that they have never been able to enjoy its beauty. If you belong to either of these categories, this book may remove some of these unnatural barriers by making the journey a pleasure.

The journey metaphor is useful: in a sense this book is a tourist guide. It describes geometric spaces containing linguistic information, and the tools we have developed to build and navigate our way around these spaces automatically, making maps as we go along. Along the way we'll encounter idioms shaped like a pair of kitchen scales, the great Tree of Life and some of its linguistic counterparts, ambiguous words which behave like semantic wormholes, and a crystalline lattice of names of horses with the same shape as the lattice derived from Aristotle's ancient theory describing the way the universe is made up of *earth, air, fire* and *water*. If you want to take a holiday in concept space from the tranquility of your favourite armchair, then make yourself comfortable and the tour will begin.

How to read this book

Different readers will be interested in different ingredients of this book, and to accommodate this diversity the main text has been supplemented in two ways. Firstly, suggestions for wider reading are included at the end of each chapter, covering a range of topics. Some articles are relevant to particular issues which were not included in the main text for fear of breaking continuity and making some paragraphs appear more like a bibliography than a book. Others are more general sources describing whole fields which affect our work, which could not be covered properly in this volume but might be of great interest and benefit to the reader.

Secondly, some material has been placed in 'special interest boxes' (Table 0.1). The hope is that these will enable those with particular backgrounds (or developing interests) in mathematics, language,[2] computing or science in general to read the material they want to, without the main text dissipating into issues which for other readers might be irrelevant or baffling detail. Many of these boxes should really belong to more than one category: this is one of many examples of the limits inherent in any attempt to classify information into a collection of non-overlapping subject groups.

Words, Quotations, Concepts and Meanings

Defining the meanings of words like *concept* and *meaning* is a great philosophical challenge which we will put to one side, relying on the reader's intuition and a few different typefaces and notational conventions to distinguish between different uses of the same string of letters.

To begin with, direct quotations such as "If music be the food of love, play on" are enclosed by double quotation marks, or placed in separate quotation paragraphs such as

> As you from crimes would pardoned be,
> Let your indulgence set me free.

There will be many examples of this kind, because these quotations often illustrate reasons why a geometric model built from a particular piece of natural language text behaves as it does.

---

[2]My thanks to Peter Kleiweg and John Nerbonne for the use of the image of signs from different alphabets used in the language boxes. (The other symbols were drawn by the author using Xfig.)

INTRODUCTION / xxi

Mathematical boxes	Language boxes
Boxes with this symbol in the corner draw the reader's attention to wider issues and stories about the mathematics we are using, and alternative viewpoints and approaches.	Boxes with this symbol in the corner contain observations about language, its uses, and some of the terminology and conventions in the study of language.
| ```
1010101
0101010
1010101
``` Computing boxes |  Science boxes |
| Boxes with this symbol in the corner give practical information about how different models and systems were built, and where to find many freely available resources for natural language processing. | Boxes with this symbol in the corner draw the reader's attention to phenomena in science as a whole, especially some which have motivated the development of the mathematical techniques we are applying to language. |

TABLE 0.1 Different threads of material catering for those with specific interests

Single quotes or 'scare quotes' are reserved for technical terms where appropriate, though there is some overlap between the use of single quotes and standard *italics*.

Sans serif italic font is used for linguistic concepts. If, for example, we say that the word *referee* is related to the word *umpire*, we usually mean that there is some representation of these concepts (dictionary definitions, regions in a geometric model, etc.) and that these representations have certain features in common.

Finally, URLs (Uniform Resource Locators, i.e. webpage addresses) and example lines of code from computer programs appear in `typewriterfont`.

I hope that the use of these notational conventions where appropriate has helped to make the examples and arguments in this book clearer.

Acknowledgments and thanks

Great thanks are due to many people who have contributed to this book in different ways. First of all, to my colleagues and students on the Infomap project, especially Beate Dorow and Scott Cederberg, who did
```

a great deal of the research and programming that went into this book, and to Professor Stanley Peters, without whom the project would never have even existed.

The book has benefited greatly from the constructive suggestions of reviewers including Tim Baldwin, Pentti Kanerva and my father Kit Widdows, and their experience, encouragement and critical insight have been invaluable.

The Infomap project received financial support from NTT (our industrial collaborators in Japan), and from the American National Science Foundation through the MuchMore project, a collaboration with partners in the European Union for multilingual medical information management. In this context, I would particularly like to thank Paul Buitelaar and Špela Vintar for great discussions and hospitality during my visits to Germany.

To Maryl, my bride and soulmate, my everlasting thanks for her love and faith in this ever-escalating project, and her enthusiastic enjoyment of the book even though she thought she hated math.

Lastly, my gratitude to many giant figures of scientific progress spanning three millenia cannot be described in a few words: hopefully some measure of my debt and appreciation is reflected in this work.

1
Geometry, Numbers and Sets

Ask someone what mathematics is about, and most of the time in this day and age something to do with numbers will be mentioned straight away (unless it's just "something I hated in high school until I was allowed to give it up"). This is a long way short of the whole story: Webster's (1997) dictionary sums this up, defining mathematics as

> 1) the science of numbers and their operations, interrelations, combinations, generalizations, and abstractions and of space configurations and their structure, measurement, transformations, and generalizations.

There are numbers, for sure, and we will find them very useful tools — but mathematics is also "the science ... of space configurations and their structure." This branch of mathematics is traditionally known as geometry, and was rigourously studied some two thousand years before the same exacting standards came to be applied to the study of numbers. The Oxford English Dictionary, with its typically traditional and didactic voice, recognizes this historical precedence, defining mathematics as

> 1. Originally: (a collective term for) geometry, arithmetic, and certain physical sciences involving geometrical reasoning, such as astronomy and optics; spec. the disciplines of the quadrivium collectively. In later use: the science of space, number, quantity, and arrangement, whose methods involve logical reasoning and usually the use of symbolic notation, and which includes geometry, arithmetic, algebra, and analysis; mathematical operations or calculations.

As this definition suggests, the traditional sciences such as astronomy and optics found geometry to be an invaluable tool, as have many other sciences, including music (understanding the relationships between notes of different pitch), mechanics (understanding the path of a projectile), chemistry and biochemistry (understanding the structure of different atoms and molecules) and physics (one of whose crowning glories, Einstein's *General Theory of Relativity*, is an entirely geometric theory). Geometry has contributed much and benefitted enormously in its interaction with these different fields. The purpose of this book is to describe the way in which a new interaction is developing between the study of geometry and the study of language — in particular, the study of meaning in the new and practical field of natural language processing. To begin this story, it is important to consider the question, "What is geometry?" Why on earth *should* the study of space configurations and their structure be beneficial when dealing with meaning in language?

As well as introducing geometry, this chapter discusses some of the relationships between geometric objects, other mathematical concepts such as numbers and sets, and the way mathematical concepts have been used to describe and model the world around us. Some of this chapter (in particular, the introductions to numbers and set theory) are partly for later reference. Any readers with little mathematical background who find these sections difficult are urged *not to get bogged down*. Much of the material in the rest of the book can be understood, albeit less formally, without going into every bit of mathematical detail. Also, some of the complicated-looking ideas and symbols introduced in this chapter may become much more familiar through being used in practical linguistic situations. Readers who follow the gist but not the details in this chapter will probably get more knowledge and *much* more enjoyment from this book by pressing on to further chapters and referring back to different sections in Chapter 1 as necessary.

This health-warning aside, the next section which introduces geometry itself is a must-read, *especially* for those who have a good grounding in formal methods generally but are unaware of geometry and its story.

1.1 What is Geometry?

We all solve many problems that can be described geometrically: it's so commonplace that you probably won't have even thought of it as 'doing geometry.' When packing the back of your car for a holiday, you arrange your bags and suitcases so that they fit within the confines of the space available. Before driving, you adjust the mirrors so that they are in just the right place so that you can see the other traffic, intuitively aware of how rotating the mirror in or out, up or down, will affect your view. You turn the steering wheel so that the car will take a safe path, avoiding other objects.

If asked to give a number saying "how far" you moved the steering wheel or the mirror (for example, if asked to measure it in degrees, 360 degrees making a complete circle), you might not have much of an idea, and it may sound like a dumb question anyway. What's important isn't how many degrees it moved: what matters is whether it moved too little, too far, or just the right amount. When packing your bags in the back of the car, the number of cubic centimetres they occupy is not so important — what's important is how their shape and size compares with the space available.

These constraints often have to be satisfied in more than one way at once. If your suitcase is longer than the storage space available in the car, then even if it's thin enough to have a very small overall volume, it still won't fit (a realistic problem when trying to fit a portfolio of art into the overhead locker on an aeroplane). The length, breadth and height of your luggage *all* have to be smaller than the available space, which is why rotating things around to match the biggest dimensions in the luggage against the biggest dimensions in the available space can be important. When adjusting the mirror, you can move it up or down, and towards or away from the car, and it has to be in the right place with regard to both of these parameters. The steering wheel only has one parameter to vary, its amount of rotation, and if you did want to use numbers, the position of the steering wheel can be described using a single number, namely the angle (positive or negative) through which it has been turned from the straight on position. In geometrical terms, getting the steering wheel into the right place is a problem in *one dimension*, positioning the mirror is a problem in *two dimensions*, and packing your luggage is a problem in *three dimensions*.[1] The steering wheel, the mirror and the luggage are all solid, three-dimensional objects, but because some of the parameters with the steering wheel and the mirror are fixed, there are fewer parameters left to vary.

---

[1] In fact, you can easily vary more than three parameters when positioning a suitcase — many simple systems can vary along many more dimensions than just two or three. Higher dimensions will be discussed properly in Chapter 5.

There are countless practical situations which use similar geometric considerations: when trying to manoeuvre a piece of furniture through a doorway, choosing the right size and shape of utensil for a task in the kitchen, what shape and size to cut your vegetables so that they end up cooked but not soggy or burnt, or opening and closing a window to adjust the temperature in a room. We *could* use numbers to solve these problems — say by measuring the doorway and the dimensions of the piece of furniture with a tape measure — but in practice we often don't. This is because we're not interested in the absolute size of the piece of furniture in some fixed external scale — what we're interested in is the *relationship* between the shape of the piece of furniture and the doorway.

Nature is also adept at developing and using geometric solutions. Many fish, such as the trout, are ideally shaped to swim through water, having evolved a similar shape to the airfoil of an aeroplane wing. Both the trout and the airfoil have their blunt end at the front and their tapered end at the back — it may be counterintuitive, but this actually causes much *less* drag than having the sharper end at the front to cut through the air or water. It is thought that moths fly in a straight line by keeping their angle with the moon constant (some even shift this angle by 15 degrees every hour to account for the movement of the moon in the sky), using this as one of their navigational techniques. This is one account for why moths end up being attracted to light: if you travel keeping a very *distant* object at a 45 degree angle to

 Language and space

Spatial metaphors appear in many commonplace linguistics expressions concerning other aspects of the world. Periods of time are decribed as *long* or *short*, events may happen be*fore* or *after*, and we even say *always* to mean "at all times." We talk about *areas* or *fields* of study, groups to the *left* or *right* in politics, and the *rise* and *fall* of all sorts of things.

More generally, cross-linguistic studies suggest that the idea of *location* plays a fundamental role in cognitive semantics, providing a grounding with- in which ideas such as possessions or experiences (*themes*) may be expressed. The correspondence between the theme and location 'case roles' and the notions of Figure and Ground in psychology is highlighted by Delancey (2000).

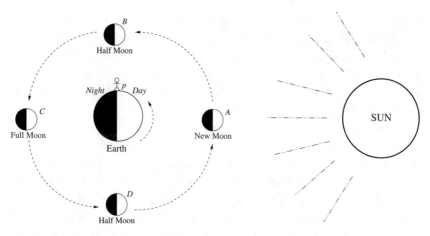

FIGURE 1.1 The sun, the earth, and the phases of the moon as seen by an observer on the surface of the earth.

your path, you will go in a fairly straight line, but if you apply the same 45 degree rule to a *nearby* object you'll spiral in towards it. (Why moths stay by the light once they get to it is more of a mystery: they may have simply not evolved a contingency plan for reaching the moon!)

People have also paid close attention to the heavenly bodies, and used simple geometric reasoning to understand what they see. For example, the phases of the moon can be understood by considering the changing angle between the sun, the earth and the moon (Figure 1.1). When the moon and the sun are at right angles (positions B and D), exactly half of the lighted portion of the moon is visible from the earth, so the observer at point p sees a half moon. When the earth lies almost directly in between the sun and the moon (position C), the entirety of the lighted portion of the moon is visible from the earth, and we see a full moon (except during a lunar eclipse, when the line-up is exact). At the other extreme, when the moon lies between the earth and the sun (position A), we see (or rather fail to see) a new moon.

This diagram can actually tell us more than this. The half moon at position B is directly overhead for the observer at p just as the sun is setting, whereas if the moon were full it would be opposite to the sun and hence would be rising just as the sun sets. In this way, the time of moonrise and moonset are easily deduced from knowing which phase the moon is in (the Ancient Greeks and Egyptians both measured the beginning of the month from the first night the new crescent moon was visible in the night sky, just after sunset). From knowing a little about the relative positions of three spheres, we have used our intuition and reasoning to predict that when the sun, the earth and the moon are in a particular spatial configuration, we will observe particular shapes in particular directions.

In this conceptual model we have made certain simplifications and approximations: we haven't taken the *latitude* of the observer into account, and the half moon will not be directly overhead for an observer who is a long way from the equator (though it will be at its highest point that night). These approximations are *inexact*, but we should hesitate to call them *wrong*. Every mathematical abstraction used as a model of the physical (or linguistic) world contains simplifications — if we tried to consider every single factor that could possibly have an effect, we would never get round to making any predictions. Rather than spending forever trying to build a model which reflects every aspect of reality, the challenge is to choose intelligent simplifications so that the most important features of the situation are clearly apparent, while being aware of the limitations these assumptions introduce. In Figure 1.1, we have reduced everything so that the earth, the moon and the sun lie in a single plane and the whole system is determined by the angle between the sun and the moon. That these assumptions are a simplification is justified by the astonishing predictive power thus released.

1.1.1 How did geometry arise?

Hopefully, the previous section gave the reader a taste of what geometrical methods can accomplish, based upon very little theory at all. This was one of the early attractions of geometry, which enabled it to forge ahead of other sciences under the ancient Greeks. In truth, our story should begin way earlier than that — in countless monuments, inscriptions, and decorations, testimony is given to the natural awareness humans have for shapes and the way they can be used to reflect and predict important events such as the changing of the seasons.

One root of geometry is attributed to the Egyptians: namely, that after the annual flooding of the Nile, fields had to be redrawn and equitably divided, a task given to professional 'rope-stretchers'. Many simple geometric constructions can be implemented on a large scale by two people stretching a rope between them — if the rope is tight, they obtain a straight line (the roots of these words are commonly quoted in English as meaning *stretched linen*), and if one person stands still holding one end of the rope and the other person walks with the other end, holding the rope tight, they obtain a circle. The word *geometry* comes from the Greek words for *earth* and *measuring*, indicating perhaps that this application was thought of as particularly important.

1.1.2 Euclidean geometry

Many of these rules and techniques were gathered together by Euclid of Alexandria, one of the most influential characters in this book (and in mathematics as a whole).

After the death of Alexander the Great (323 BC), his huge empire which stretched from Greece to India was carved up between his generals. Ptolemy, Alexander's close friend and commander of cavalry, gained control of Egypt and founded a dynasty of 'Greek pharaohs' which lasted until the death of Cleopatra (30 BC), when Egypt finally became a Roman province. Ptolemy, a great patron of learning, founded the University of Alexandria which became arguably the greatest centre of scholarship in the Ancient World, and among the first generation of teachers who Ptolemy attracted to his University was Euclid.

The textbook written by Euclid, his *Elements of Geometry*, was first used around 300 BC and became by far the most widespread

> **Euclid on the Web**
>
> Euclid's *Elements*, written some 2250 years before computers were invented, is the perfect electronic book. The very best way to get an introduction to Euclid is to use the online version of David Joyce (Clark University), available at `aleph0.clarku.edu/~djoyce/java/elements/elements.html`, or by typing "Euclid Elements" into most internet search engines.
>
> Hyperlinks enable the reader to quickly call upon previous axioms and propositions, and there are many live Java applets where the reader can drag different points around, seeing before their eyes the effect this has on the points around them, demonstrating that the constructions are valid for *any* points, not just a few specially chosen ones.

book on mathematics throughout space and time, used to teach young people geometry until its gradual replacement in schools over the past hundred years or so (at great cost to the welfare of students, some would still argue).

Almost all of the results in the *Elements* are traced to mathematicians before Euclid's time, particularly Pythagoras (ca. 580–500 BC), who we shall meet again shortly, and Eudoxus (ca. 400–340 BC), a prominent member of Plato's Academy. Euclid's genius lay in the way he systematically developed the results of geometry (called *Propositions* or *Theorems*) from a very few assumptions called *Axioms*. Some of the axioms are special to geometry, the first axiom stating that

1. Between any two points, a straight line can be drawn.

the third that

3. A circle can be drawn with any centre and radius.

Words for space and place

Classical Greek, like modern English, had different words for different notions of space. *Topos* referred to the place occupied by a physical body. *Chora* referred to a local region, such as the hinterland of a particular city. *Kenon* referred to empty space, which could potentially but need not be occupied by an object. *Place*, *room* and *void* are often proposed as translations for these three concepts.

The kind of space we conceive of affects the rules we intuitively apply when reasoning: for example, two physical objects cannot occupy the same place (i.e. *topos*), but two people can certainly work in the same place (i.e. *chora*).

and so on: there are five geometric axioms of this sort.[2] There are also five general axioms which Euclid calls *common notions* — that is, they are common to the whole of mathematics. These include

1. Things which equal the same thing also equal one another.

2. If equals are added to equals then the sums are equal.

and so on. The rest of the propositions of geometry can be deduced from these humble beginnings. The first proposition is shown in Figure 1.2,

---

[2]These axioms can sometimes be thought of as 'ruler and compass' constructions — the first says that if you have a long enough ruler, you can draw a straight line between any two points, and the third states that if you have a big enough pair of compasses you can draw a circle of any size, with any point at its center.

which proves that an equilateral triangle can be constructed on any straight line by drawing two circles. You may not feel that this in itself is a groundbreaking achievement: however, by the time Pythagoras' theorem (see page 102) has been proved from these same axioms (in Euclid's Book I, Proposition 47), you may begin to feel quite impressed.

This approach has become the very cornerstone of the way mathematics is developed and studied: we make a few assumptions that appear to be generally true, and based upon these we see what conclusions follow. In any situation, or for any structure where the original assumptions are shown to be true, it follows that all of the conclusions are true as well. Many rich and sophisticated mathematical practices were developed in Egypt, Mesopotamia, India and China and it would be utterly mistaken to say that the Greeks 'invented' mathematics: however, it is probably fair to say that Greek Geometry set new standards for *proof* in mathematics, transforming well-known rules-of-thumb into an exact science.

In many ways, this book tries to take a similar approach to learning about language. For example, in Chapter 2 we use a simple assumption that two nouns which occur together in a particular pattern are similar in meaning, program a computer to find all such pairs of nouns in an enormous body of text, and explore the resulting space of nouns and links to similar nouns. Such simple axioms do a much less reliable job of describing language than describing straight lines, because human language contains so many exceptions to nearly every rule: but if we tread carefully, the general method proves to be very powerful.

1.1.3 Individual objects and relationships

As well as the Axioms of Geometry and the Common Notions, Euclid's work also begins with 23 definitions, such as

1. A *point* is that which has no parts.

and

2. A *line* is breadthless length.

Euclid also defines notions such as *circle* and *equilateral triangle* in his introduction. It has been argued that some of Euclid's definitions are far from satisfactory — after all, what does it mean to say that a point is something that doesn't have any parts? Maybe this statement would make more sense in its historical context, where Greek philosophers

Elements of Geometry, Book 1, Proposition 1

To construct an equilateral triangle on a given finite straight line

Let AB be the given straight line.
It is required to construct an equilateral triangle on the straight line AB.

Describe the circle BCD with centre A and radius AB.   *(Axiom 3)*
Again describe the circle ACE with centre B and radius BA.

Join the straight lines CA and CB from the point C at which the circles cut one another to the points A and B.   *(Axiom 1)*

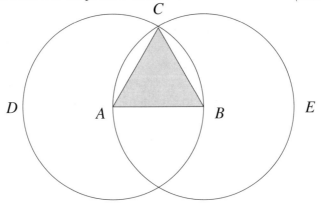

Since the point A is the centre of the circle CDB, AC equals AB.
Again, since the point B is the centre of the circle CAE, BC equals BA.

Therefore each of the straight lines AC and BC is proved equal to AB. And things which equal the same thing also equal one another, therefore AC also equals BC.   *(Common Notion 1)*

Therefore the three straight lines AC, AB and BC equal one another.

Therefore the triangle ABC is equilateral, and it has been constructed on the straight line AB as desired.

FIGURE 1.2 Euclid's first proposition, which can be deduced from two of the axioms of geometry and one of the common notions.

such as Leucippus and Democritus had argued that matter consisted of indivisible *atoms*, and debated the nature of these atoms — it is possible that Euclid felt obliged to make some basic statements about what sort of points he was talking about.

Here we begin to encounter the divergence that occurred around the time of Euclid, between the concerns of mathematics and those of philosophy. To this day, mathematicians are largely unconcerned with asking "what are mathematical objects?" in spite of the profound philosophical troubles that might ensue, and when philosophers draw attention to these problems, mathematicians tend to ignore such things with a shrug and get on with doing mathematics. (In a sense, we all do this with language — whatever problems there are in giving a proper account of the philosophy of language, we all continue using language very effectively for both basic and profound communication.)

Rather than trying to define what objects like points are, mathematicians, and geometers in particular, are much more concerned with describing the *relationships* between them. Euclid's axioms and postulates are a paradigm example of this. The most significant axioms are all about relationships: that is, they are all about statements which describe or link more than one point. Take the first axiom of Euclid's geometry: *between any two points, a straight line can be drawn*. This describes a clear relationship between two points and those that lie on a line joining them. Or the first common notion: *things that equal the same thing also equal one another*. This says that if the relationship $A = B$ holds and the relationship $B = C$ holds, then the relationship $A = C$ also holds.

This long tradition of studying objects by studying first and foremost the relationships they are involved in lies at the heart of this book. We will study relationships between words and concepts, how they can be learned by a computer just by reading normal language, and how many such relationships can be combined into a whole structure of knowledge. Individual relationships between pairs of concepts might include

A *car* is a kind of *vehicle*.   A *steering wheel* is part of a *car*.

or possibly

> A *car* is quite similar to a *van*.

None of these on its own tells us the whole story about the *car*, but once you put them all together and start following the paths that these relationships take us along (for example, by next asking "All right, so what can you tell me about a vehicle?") a lot of progress can be made. One way to describe this approach is to talk about a *network* of concepts (Quillian, 1968): some of its aspects could be described as *connectionist* or *associative* (Gärdenfors, 2000, p. 1). The point to remember is that in Euclidean geometry, and in our study of the geometric structures that can be used to represent meaning, it is the *relationships between points* that are the key.

Object-Oriented Databases

Object-oriented databases are often interconnected systems where much of the information contained by different objects is represented by their links to other objects. For example, in a bibliographic database, each author, each publisher, and each book might be an object. These objects contain links to one another, and using this mechanism queries like "find me the author of book x, and find me the titles of other books by this author" can be supported. Database designs are often presented in pictorial diagrams to make their network or 'graph' structure clear.

The way in which objects are linked, the length of a path from one object to another, and the design of indexes which aggregate link information from different parts of the network, are all factors to in making such a database meaningful and usable (Bertino et al., 1997, Ch. 1).

1.2 Sets, Relations and Mappings

The concept of a *set* is one of the most fundamental and flexible notions in modern mathematics. (Indeed, the word *set* is one of the most flexible and ambiguous words in English.) The concepts, and particularly the notation of set theory will be used throughout this book, so it is as well to give a brief summary. Unfortunately, like any new notation, the language of set theory takes some getting used to, and it must be admitted that the reader who is unfamiliar with statements such as $a \in A$, $A \subseteq B$ and $f : A \to B$ will struggle with the formal equations in this book. In which case, do not despair: if something really looks like Double Dutch, you are completely within your rights to skip it, and if the main point isn't made clear with a diagram, a description, or an example, then the fault is mine.

A *set* is, broadly speaking, any collection of distinct objects or things, be they real or abstract. The collection of books on my desk forms a set, the collection of people in the world forms a set, and the points on the surface of the moon form a set. If an object p is in a set A, we write $p \in A$, and we call p a *member* or an *element* of A. For example, if we let M be the symbol for the set of all points on the surface of the moon, then $p \in M$ means "p is a point on the surface of the moon."

The simplest way to define a set is to list its members, which is conventionally done within curly brackets. For example the set C of primary colours could be defined by listing

$$C = \{\textit{red, green, blue}\}.$$

An ellipsis or series of dots can be used to imply that you have missed out some of the elements of the set that the reader is meant to assume are included: for example, an important set in mathematics is the *natural numbers* which may be written as the set

$$\mathbb{N} = \{1, 2, 3, 4, \ldots\},$$

where the effect of the row of dots is very similar to that of writing "etc." in text.

Two sets A and B are *equal* if they have the same elements, so every member of A is also a member of B, and vice versa. This equivalence is sometimes called the *Axiom of Extensionality*, and we will see why later in this book.

If a set A is contained in a larger set B, we say that A is a *subset* of B and write this relation with the symbol $\subset$ so that $A \subset B$. For example, the set of children is contained in the set of people, and this could be written

$$\{\textit{children}\} \subset \{\textit{people}\},$$

which also corresponds to the statement "children are people." The slightly looser symbol $\subseteq$ in the statement "$A \subseteq B$" means "A is a subset of B or the same set as B," or (equivalently and very importantly) "B contains A," so in symbols

$$A \subseteq B \quad \text{if and only if} \quad A \subset B \text{ or } A = B.$$

14 / Geometry and Meaning

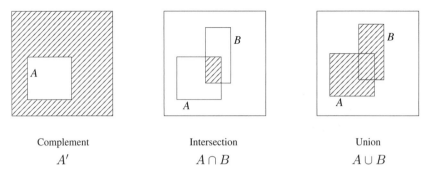

Complement	Intersection	Union
A'	$A \cap B$	$A \cup B$

FIGURE 1.3 The shaded regions represent the set complement A', intersection $A \cap B$, and union $A \cup B$.

This corresponds loosely with the relationships 'greater than' and 'equal to' for numbers. Recall that, for two numbers x and y, we write $x < y$ to mean "x is smaller than y." The symbol $\leq$ means "is less than or equal to," so for two numbers x and y,

$$x \leq y \quad \text{if and only if} \quad x < y \text{ or } x = y.$$

The analogy between the $\subset$ relation on sets and the $<$ relation on numbers is made much clearer in Chapter 8, where we discuss *order* relationships.

As well as these relationships between sets, there are several important ways of getting new sets from old, three of which are shown in Figure 1.3.

The *complement* of a set A consists of everything in the universe *except* A. The complement of a set A is often written A'.

The *intersection* of two sets is the set of objects which are in both of them, and is represented using the symbol $\cap$. For example, if D is the set of all **drummers** and B is the set of **Beatles**, then their intersection is the set[3]

$$D \cap B = \{\textbf{\textit{Ringo}}\}.$$

Some familiar mathematical statements can easily be written down as intersections of sets — for example, "the only even prime number is 2" can be written something like

$$\{\text{even numbers}\} \cap \{\text{prime numbers}\} = \{2\}.$$

---

[3]This simplification is accurate for the most part, though Pete Best and Paul McCartney also played the drums with this group (see Harry, 1992, p. 58).

The *union* of two sets is the set of objects which are in *either* (or both) of them, and is represented by the symbol ∪, which is often remembered as being "U for Union." The word *union* or *united* is often used with roughly this meaning in normal or official language — for example, the official name of the UK is "The United Kingdom of Great[4] Britain and Northern Ireland" and this definition could be written mathematically as

$$\{\text{United Kingdom}\} = \{\text{Great Britain}\} \cup \{\text{Northern Ireland}\}.$$

In a sense (which will be made clear in Chapter 8), union can be thought of as a partner operation of intersection, and their symbols were chosen to make this relationship clear.

Set theory is a very general way of expressing mathematical concepts and the relationships between them. For example, back in the model of the sun, earth and moon of Figure 1.1, let us call the set of points on the surface of the moon M. Then the lighted part of the moon is the portion of the moon that is visible from the sun, or

$$lighted(M) = \{p \in M \text{ such that } p \text{ is visible from the sun}\},$$

so the set $lighted(M)$ is a subset of the surface of the moon M. The shape that is actually visible in the sky, which we shall call $crescent(M)$, is the subset of $lighted(M)$ that is visible from the earth, or

$$crescent(M) = \{p \in lighted(M) \text{ such that } p \text{ is visible from the earth}\}.$$

We could perform both of these steps at once, saying that

$$crescent(M) = \{p \in M \text{ such that } p \text{ is visible from the sun}$$
$$\text{and } p \text{ is visible from the earth}\}.$$

In fact, we could define the subset $earthview(M)$ to be the set of points on the moon which are visible from the earth, in which case the set $crescent(M)$ consists of precisely those points which are in both $lighted(M)$ and $earthview(M)$, which would be written

$$crescent(M) = lighted(M) \cap earthview(M).$$

In this way, the different regions of the moon we are interested in can be expressed mathematically in terms of a very few sets and

---

[4]The word *great* was added not to signify grandeur but to distinguish larger Britain from smaller Britain, which is Brittany in northwest France. The word *great* appears with this meaning in many geographical names such as 'Great Gable' and birdnames such as 'great tit.'

relationships between points in those sets. This combination of brevity and concreteness is the motivation for using set theoretic concepts and notations, though some readers may find this a small reward for having to wade through all the symbols in this section. (Many of these symbols, including the ∩ for intersection and ⊂ for "is a subset of" are said to have been introduced by Peano as he developed the axioms for the natural numbers.)

1.2.1 Containment and Implication

One important application of set theory arises because of the relationship between the geometric idea of *containment* and the logical idea of *implication*, which is one of the fundamental connections between mathematics and the study of meaning. We will encounter this principle over and over again throughout this book, and since the connection is not difficult to grasp, it is worth drawing attention to it early on.

Consider the following two assertions:

If Argos is a dog, then Argos is an animal.

The set of dogs is contained in the set of animals.

Both of these statements follow from the same basic premise, which is that dogs are animals. The left-hand version describes this in terms of the logical relation between two assertions ("Argos is a dog" and "Argos is an animal"), whereby if the first assertion holds then the second one also holds (which would remain the case if we swapped the name Argos for any other name, or just any algebraic symbol like A or x). The right-hand version describes the premise in terms of the geometric relation of containment between the set of dogs and the set of animals.

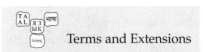
Terms and Extensions

Let a be a concept. Its *extension* $\text{Ext}(a)$ is the set of objects which are referred to by a — for example, the extension of the concept *dog* is simply the set of all dogs. The difference between a concept and its extension can be much less trivial than this — for example, it happens that "the Head of the Commonwealth" and "the Patron of the Kennel Club" refer to the same person (Queen Elizabeth II), so the *extensions* of these two concepts are identical, although we would probably agree that their *meanings* are quite different.

These statements can easily be written using logical and set-theoretic symbols, as follows:

Logical
A is a dog $\Rightarrow$ A is an animal

Set-theoretic
$\{\text{dogs}\} \subseteq \{\text{animals}\}$

We use this equivalence to make very simple inferences about locations — for example, if Jacque is in Paris then Jacque is in France, which follows because we know that Paris is contained in France. (This presumes, of course, that we are talking about Paris, France rather than Paris, Texas or any other Paris. This sort of ambiguity will be treated in Chapter 3.)

The equivalence between containment and implication is recognized clearly by Aristotle (once one gets used to terms such as *predication*), very near the beginning of his foundational treatise on logic:

> That one term should be included in another as in a whole is the same as for the other to be predicated of all of the first.
>
> (*Prior Analytics*, Bk I Ch. 1)

The somewhat tortuous use of words like "other" and "first" in this description to refer to the different terms makes the use of symbols such as A and B very tempting, and in his very next chapter Aristotle introduces precisely this innovation, which gradually revolutionized the way mathematics and science is written and explained.

1.2.2 Mappings between sets

The final mathematical concept we will define in this section is a *mapping* between sets. A mapping f from a set A to a set B is a rule which associates a member of B with each member of A. The name mapping is an analogy from cartography, in which a map is a picture and each point on the picture corresponds to a point on the earth's surface (or whichever terrain is being represented in the map).

A simple mapping in day-to-day life is the pricing of goods in a catalogue, which is a mapping from the different items which can be purchased to a number. Each item p is mapped to a number $f(p)$, the *price* or *cost* of p. The *range* of possible prices is usually a discrete class of numbers consisting of a whole number of units (dollars, Euros, pounds, etc.) and a whole number of hundredths, often called cents.

The process of adding 1 to a natural number n can also be thought of as a mapping, from the set of natural numbers (denoted by the symbol $\mathbb{N}$) to the set of natural numbers. That is, we can define a map

$$f : \mathbb{N} \longrightarrow \mathbb{N}$$

by stating that, for every natural number $n \in \mathbb{N}$, $f(n) = n+1$, or "f maps n to $n + 1$." This may seem like a very fussy and confusing way to use a lot of symbols to define something you've been able to do ever since you learnt to count. The power of such ideas is that they enable you to use the same techniques and the same language to talk about situations that appear to be completely different — such as adding whole numbers, the pricing of goods in a catalogue, and the shape of the moon you see in the sky.

This is one of the great strengths of abstract approaches. Often when challenged that mathematics is too abstract to be really useful, I've tried to explain that its very abstraction enables mathematics to be useful in such a great variety of situations. The great challenge is in finding *which* mathematical abstractions are appropriate tools for describing the situation you're working with, and in recognizing where they fall short. It is crucial to be aware that mathematical models always *do* fall short of giving a complete description of any real world situation, not because they are so complicated but for the opposite reason: they make so many simplifications.

1.3 Numbers and Ratios, Mysticism and Music

Numbers are another cornerstone of mathematics. As we've seen, there is a great body of interesting mathematics that doesn't depend on numbers: however, as soon as we need to count and measure different events, numbers are invaluable. Over the centuries, numbers have been used to represent many different concepts and have often assumed a life of their own, partly because special objects and relationships have had special numbers associated with them.

1.3.1 Numbers and Elements

The oldest, simplest and most primitive use of numbers is the first one we learn as children: counting. With the advent of set theory, it became easier to give a formal definition of what this means: counting is a process of measuring the size of a set, and a number in one of its most fundamental senses (called a *cardinal* number in mathematics) refers to

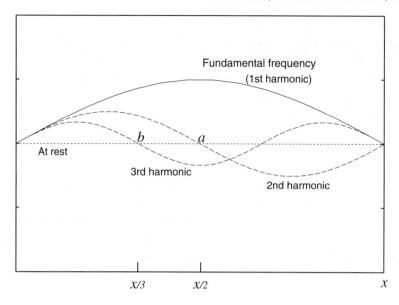

FIGURE 1.4 Harmonic waves on a string

a measure of how many elements there are in a given set. However, this presumes that objects can be enumerated separately, and in spaces with infinitely many points, this assumption breaks down and other ways of measuring and comparing must be used. One way is to count the number of *dimensions* of a space, which allows us to compare the number of primitive or independent elements out of which infinitely many new points or objects can be constructed. The mathematics behind these processes will be introduced gradually.

The use of the term *elements* to describe the primitive substances from which a whole space is created is a deliberate allusion to ancient traditions, which have relied on our understanding of the natural world, and the mythological traditions surrounding different numbers, to create philosophical descriptions of reality and its basic causes. The number of basic elements and the relationships between them has been used to explain all sorts of patterns, and to link ideas in geometry, religion, music, and science.

In music, for example, there is a fundamental relationship between the pitch of the notes made by a plucked string, and the length of the string. If you divide the length of a string on a stringed instrument (such as a violin, piano, or guitar) exactly in half, you will generate a note that sounds higher in pitch, but curiously sounds as if it is the *same* note. In Western music, we say that the sound generated by the halved string is one *octave* higher, though this term presupposes that the intervening scale is divided into 8 notes.

Halving the length of a string will produce a note one octave higher: dividing it in three will produce a note one octave and one fifth (12 notes) higher, etc. These different but closely interrelated waves are called the *harmonics* of the string. The first three harmonics of a vibrating string are shown in Figure 1.4. The second and third harmonics can easily be produced on a guitar string by placing your finger very gently on the string exactly one half or one third of the way along and plucking the string with your other hand. (These points are marked by the letters a or b in Figure 1.4 and they are directly above the 12th fret and 7th fret on a guitar fingerboard). This ensures that this point on the string cannot move, so only the waves which do not disturb this point can be transmitted — the higher harmonics.

These ratios between different pitches are crucially important to musical systems all over the world. A systematic and fascinating survey of this topic, and the common roots of different traditions in numerical relationships (at least in China, India and Europe) is given by Daniélou (1943). Westerners attribute the discovery of the crucial ratio 2:1 between frequencies an octave apart to Pythagoras (Daniélou's book deliberately attacks this ignorant tradition). The use of the word **octave** to describe this interval signifies an important choice in Western music, where this interval of 2 : 1 is broken down into 8 distinct pitches.

The Chinese system, on the other hand, is generated by the ratio 2 : 1 (between the first and second harmonics) and the ratio 3 : 2 (between the second and third harmonics). This ratio of 3 : 2 is also crucial in musical systems: it is called a *fifth* in Western music and is the interval between the notes C and G. Applying this ratio five times gives five distinct pitches (which roughly correspond to the notes C, G, D, A and E in Western music), and this gives the characteristic *pentatonic scale* of traditional Chinese music. There are many other significant groups of five, including the visible planets (Mercury, Venus, Mars, Jupiter and Saturn), whose movements Chinese astronomers could predict with

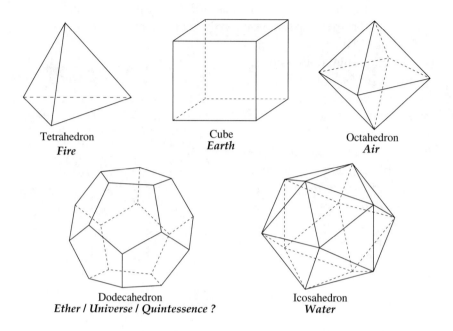

FIGURE 1.5 The Platonic solids and their associated elements

great accuracy, and the five *elements* of Chinese philosophy (earth, water, wood, metal and fire).

The traditional number of elements in Greek philosophy was not five but four — earth, air, fire and water.[5] Many traditions agree that the first scientist to describe *all* of these elements was Empedocles (ca. 490–430 BC) (see page 268). Plato (427–347 BC) also described the creation in terms of these four elements in his *Timaeus* dialogue, and to each one he associated one of the regular solids which are called the *Platonic solids* to this day. The Platonic solids are the regular 3-dimensional shapes, whose faces are all equal and regular polygons, and during his youth Plato was only familiar with the cube (made out of 6 squares) and the tetrahedron, octahedron and icosahedron (made out of 4, 8 and 20 triangles respectively). The association between the tetrahedron and the

---

[5]This set of four elements is also used in various astrological traditions, which may suggest that it is much older than the early Greek philosophers.

element of fire goes back even further — the name *pyramid* means 'flame shaped' and is from the same root as the word *pyromaniac*.

During Plato's time, a *fifth* solid called the dodecahedron (made out of 12 pentagons) was discovered, possibly by a member of Plato's academy.[6] From around this time, a fifth element was also discussed by philosophers, sometimes associated with the *ether* or quintessential stuff of the universe (*quintessential* means 'of the fifth essence'). Not for the last time, a new mathematical possibility was discovered and the physical sciences tried to fit it into the system. This process continues to this day: for example, because of gaps in the periodic table, some of the chemical elements were predicted before they were actually discovered. Experiments in particle accelerators have been designed specifically to try and detect new subatomic particles whose properties such as mass, charge and spin are predicted by mathematical symmetries.

 Number relationships

As well as certain numbers being accorded special significance, there are many systems where the relationship between two numbers is used to describe a particular tendency or attraction. This is often the case in music: for example, in the Western major scale of 8 notes, a tune which alights on the 7th note has a strong tendency to complete the octave by moving up to the 8th, so much so that the 7th note is called the *leading note*. This is different in the blues scale, where the gap between the 7th and 8th note is widened, giving the 7th note greater independence.

Curiously, this relationship is echoed in chemistry, where an atom with 7 electrons in its outer shell (such as chlorine) is dangerously reactive, because it only needs one more electron to have a stable outer shell of 8 electrons.

Another attempt to link the Platonic solids with the physical sciences was made through the study of planetary orbits — for example, the Renaissance astronomer Johannes Kepler (1571–1630) found a striking correspondence between the radii of the Platonic solids (nested inside one another like Russian dolls) and the distances of the planets from the sun. Since each of the 5 Platonic solids was supposed to describe a ratio

---

[6]Finally, Euclid himself put an end to this list of discoveries by demonstrating (in the 13th book of the *Elements*) that there can be no other regular solids, one of the greatest yet simplest mathematical proofs of ancient times.

between two of these distances (Mercury to Venus, Venus to Earth, etc.), Kepler could explain why there were only 6 planets, an inference which failed rather dramatically when William Herschel discovered the planet Uranus in 1781! During his own lifetime, however, Kepler realized that the model did not correspond closely enough to observed reality, and instead of clinging to the Platonic solids, he developed three accurate laws of planetary motion which set the stage for Newton's great theory of gravitation.

The Platonic solids have proved more effective on a microscopic level, because many crystalline structures are accurately described by these shapes. For better or worse, the Platonic solids have encouraged some of the most elegant, the most persistent, and sometimes the most unfounded attempts to combine geometry with meaning.

Numbers throughout the ages have been used for much more than just counting: they are used to encode and express deep ideas of stability and change in many different walks of life. However, this history is both an inspiration and a warning to be on our guard against spurious correlations when trying to develop geometric models for something as intricate as human language.

1.4 Numbers and Measurement

Numbers have been used to forge many successful and elegant links (and many spurious links) between different aspects of reality. They have also been used in a very systematic and reliable fashion to measure and compare quantities, and this will be the main use of numbers in this book.

1.4.1 The Babylonians

The Babylonians are widely regarded as having the most effective system (and theory) of numbers in antiquity. It was based upon the sexagesimal system (base 60) which still influences our units of degrees, minutes and seconds — a very wise choice, because it makes the 2, 3, 4, 5 and 6 times tables as easy to learn as the 2 and 5 times tables are in decimal notation. Unlike the Greeks, Romans and Egyptians, the Babylonians also used a form of *positional notation*. This very important advance allows one to represent *different numbers using the same symbols*. This is so familiar to us that we take it for granted — for example, we know instinctively that the symbol 12 means "ten plus two" not "one plus two."

The benefits of such a system are not so striking when carving ceremonial inscriptions on temple walls, but become an innovation of great significance when doing practical calculations. For example, compare the difficulty of working out the long multiplication

$$23 \times 16 = 368$$

in Arabic numerals with the corresponding equation in Roman numerals,

$$\text{XXIII} \times \text{XVI} = \text{CCCLXVIII}.$$

The Babylonians made this advance some 4000 years ago, about 1500 years before the development of Greek mathematics. The positional number system was also used for fractions, enabling the Babylonians to calculate fractions with an ease which was not enjoyed by Westerners until the Flemish mathematician Simon Stevin and the Scottish Baron John Napier popularized the use of the decimal point (or decimal comma) in the late 1500s and early 1600s, three and a half millenia thereafter.

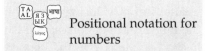

Positional notation for numbers

The system of positional notation, where a number symbol refers to a larger or smaller quantity depending on *where* it is written with respect to its neighbours, is of considerable linguistic as well as mathematical significance. It is probably the first example of a system of writing where a single symbol (such as 3) could refer to many different concepts (such as *three, thirty, three hundred*, etc.), along with systematic conventions so that the correct meaning of the symbol in different contexts could be readily inferred. For the sake of efficiency, the symbol *3* is deliberately allowed to be ambiguous, because the writer and the reader have agreed on a common format whereby this ambiguity can always be correctly resolved.

1.4.2 Greek number theory

Unlike the Babylonians, the Ancient Greeks never developed a positional notation for numbers, and were hampered by a system of using some letters from their alphabet for units, some for tens, and so on. But as well as this, Greek number theory was shattered in its infancy by a scandal which was not entirely resolved until the 19th century.

Greek number theory began, and received its greatest setback, with the Pythagoreans. Elated by the knowledge that whole number ratios caused musical notes to sound harmonious when played together, they began to develop a whole theory of numbers based on the ratios of whole numbers: what we would nowadays call *fractions* or *rational numbers*. They began to suspect that every possible relationship between two lengths could be expressed in this way: that is, for any two rods A and B, whole numbers x and y could be found so that the ratios

$$\text{length}(A) : \text{length}(B) \quad \text{and} \quad x : y$$

were the same. They soon found that this is not the case using their own famous theorem — in a right-angled triangle where the two shorter sides have a length of one unit, the hypotenuse has length $\sqrt{1^2 + 1^2} = \sqrt{2}$. The Pythagoreans sought in vain for two whole numbers which would give this (obviously important) ratio (Figure 1.6). It turns out that *no such whole numbers exist*, as can easily be proved (Boyer and Merzbach, 1991, p. 73).

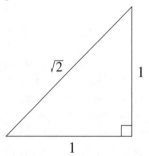

FIGURE 1.6 The hypotenuse of this triangle has an irrational length

This discovery led eventually to the distinction between *rational* and *irrational* numbers, but at the time the hole in the theory stunned the Greek mathematicians: there seemed to be no adequate theory of what 'numbers' could be if ill-behaved cases like this arose. (There is even a legend that the Pythagoreans swore to suppress this dangerous discovery, and one of their brotherhood, Hippasus, was drowned for spilling the beans.)

To some extent, Plato's collaborator Eudoxus (ca. 395–340 BC) succeeded in developing a theory of ratio or proportion which solved the worst of the problems, and allowed Greek mathematicians at least to compare the relative sizes of (for example) squares and circles (*Elements*, Book XII, Prop. 2), even though they had no way of representing a concept such as π in the way we do: using only ruler and compass to construct a square and a circle of equal area turns out to be a theoretical impossibility (which is why we use the idiom "squaring the circle" for an activity which is trying to force something impossible to happen). Euclid also made contributions to the theory of *whole* numbers, including the famous theorem that the set of prime numbers is infinite (*Elements*, Book IX, Prop. 20). However, the Greeks focussed increasingly on geometry, where the rules and foundations were felt to be much more secure, and the study of numbers faded into the background for centuries.

Nonetheless, numbers had been used in a very important way. By choosing an arbitrary length as a *unit* (such as a foot or a metre or the wavelength of sodium light), other lengths could also be expressed as numbers, by giving their *ratio* or *relationship* to this unit length, as in Eudoxus' theory of proportion. The fact that these ratios are not all whole numbers does not invalidate this process: it lies at the heart of science, in the concept of *measurement*.

1.4.3 Measurement

The word *measure* has many meanings, systematically related in many elegant ways to one another and to other concepts such as *moderation*, *medicine* and *meditation*.[7] In our specific context, *measurement* refers to making some scientific observation of a system, and in particular, giving a *number* to that observation.

This idea is as ancient as it is natural. The simplest numerical process of *counting* is a way of giving a whole number measurement to a collection of discrete objects. The measurement of *time* is particularly tempting, because different naturally occurring lengths of time (such as the daily alternation of light and darkness and the yearly cycle of the sun's position and warming effect) are readily comparable. Most every culture has recognized that the ratio of these lengths of time is about $365 : 1$, and the Egyptians are credited with being the first people to make the more accurate measurement of $365\frac{1}{4} : 1$. (This extra quarter is the reason for having a 366-day leap year in our calendar every fourth year, so as to balance the books.) The benefits of knowing this ratio for deciding when to plant crops are obvious (it's much easier to deduce that "we should sow now so that we can reap in 120 days time" and get

---

[7] A beautiful account of the concepts of measurement, ratio and wholeness in Eastern and Western traditions is given by the quantum physicist and philosopher David Bohm (1980, pp. 25–33).

the time exactly right, rather than to reckon that "we should sow now to reap after one third of a year"), and from these beginnings the various calendars developed. It's also beneficial to be able to measure spatial quantities, though the natural world gives a less compelling basis for measurement — to this day, different systems (for example, metric and imperial) are used in different parts of the world, whereas if any culture had a system of time-units that wasn't easily commensurable with the length of a day, we would probably find it absurd.

Ascribing a number to a physical quantity through measurement is thus intricately bound to a choice of *units* or a *scale*. This choice is made by choosing a mapping from some set of numbers to the system we are observing, so that each number will conventionally represent a unique *state* of the system. For example, in our model of the earth, the moon and the sun, we incorporate measurement into the system by mapping the numbers from 0 to 360 evenly onto a circle around the earth (Figure 1.7). The point on this circle which lies on a straight line between the earth and the moon is then called the *angle* (in degrees) between the sun and the moon, as observed from the earth. Note that the numbers 0 and 360 are mapped to the same point on the circle — these angles therefore represent the same state of the system, as do many other pairs such as 270 and -90. In linguistic terms, we might say that these numbers, when used to measure angles in degrees, are synonyms of one another.

The technique of measurement allows us to make much more accurate statements than just using comparisons. For example, by measuring the angle between the moon and the sun in degrees, and comparing this with a measurement system for time in (say) hours and minutes, we can say not just that "the crescent moon will be visible just after sunset," but we can predict the amount of time that will elapse (measured in hours and minutes on a clock) before the crescent moon sets.

1.5 Different kinds of numbers and measurements

Different practical situations require different sorts of numbers to make measurements. Quantities where the difference between two measurements of different states may be arbitrarily small are made using *continuous* variables. Quantities measured using continuous variables usually include length, time, angles and temperatures, among

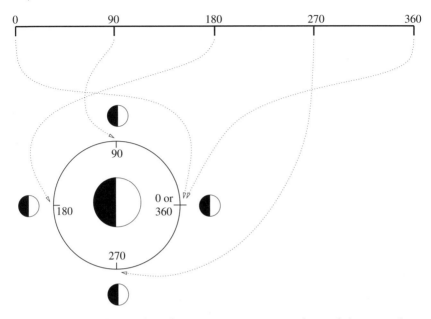

FIGURE 1.7 Using the numbers from 0 to 360 to measure the angle between the sun and the moon.

others. In a sense, there is no limit to the degree of accuracy with which these quantities can be measured, so they demand a number system which is sensitive to these arbitrarily small variations.

1.5.1 The Real Numbers $\mathbb{R}$

The Pythagoreans originally believed that fractions or *rational numbers* were suitably fine-grained to measure all possible quantities, but we know that this isn't the case — in theory, you can't measure the hypotenuse of the triangle in Figure 1.6 using fractions alone, because $\sqrt{2}$ isn't a ratio of whole numbers. This is a deep conceptual problem — in real life, the moon does not seem to skip any of its possible positions on its journey round the earth, so we need a system of numbers which doesn't have any of these gaps. Two lengths whose ratio could not be written as a ratio of whole numbers were called *incommensurable* by the Greeks — they could not be measured together using the numbers that were available to the Greeks, whatever length you chose to represent one unit.

The general set of numbers which is all encompassing enough to measure the length of *all* possible lines is called the *real numbers*, and this set is given the symbol ℝ. A real number can be thought of as an unboundedly long decimal fraction — the conceptual limit of the process of getting an ever better approximation by representing the number using more and more decimal places. The geometric significance of the real numbers for modelling all possible measurements of lengths is made very clear in language — mathematicians often refer to the real numbers ℝ as "the real *line*."

Sometimes we only use a subset of the real line to make measurements. When measuring lengths or times, we usually use only positive numbers, though we might use negative numbers if we want to keep track of which direction we're going along a line as well as how far we've gone. For systems which cycle round and round, we use an interval of numbers (from 0 to 360 in the case of measuring angles in degrees; from 0 to 12 or 24 when measuring time in hours) and when we get to the end we cycle back to 0.

1.5.2 Fixed points of reference

All of these measurements, though presented as measurements of individual objects, such as the position of the moon or the time of a particular event, are still measurements of relationships, though this is often disguised. In order to make this measurement at all, we need to decide — naturally or arbitrarily — which state to measure things *from*. In the case of the moon, we decided that the state where the moon and the sun are in the same direction from the earth should be given the number zero as its measurement. (This is built into the very idea of an angle, which by definition is a relationship between two lines, or three points — so when measuring the angle between the sun and the moon, as seen from the earth, it would seem very strange to choose any other state as zero.)

When measuring time, we also have to choose a point of reference. Choices have varied throughout cultures and periods — one typical way was to take the accession of certain rulers as reference points ("In the 17th year of our glorious Emperor so-and-so"), an ideal way for a historian to flatter a patron but a drawback when trying to compare the order of historical events more generally. The convention of measuring historical events by counting the number of years before

TABLE 1.1 Physical properties and scales commonly used for their measurement.

Property	Units	Zero point
Altitude	Feet / metres	Average sea level
Speed	Miles/km per hour	Rest
Latitude	Degrees	Equator
Longitude	Degrees	Greenwich meridian
Age (adults)	Years	Birth
Age (babies)	Months	Birth
Temperature	Kelvin	Absolute zero
Temperature	Celsius	Melting point of ice
Temperature	Fahrenheit	Melting point of mixture of half ice, half salt

or after the birth of Jesus of Nazareth (ca. 6 BC–30 AD) is attributed to Dionysius Exiguus, a 6th century monk, and for historical reasons this measurement has become widely accepted, even though it is generally believed that Dionysius miscalculated.[8] Note that there is no 'year zero' in this scheme, possibly because there was no 'zero' in Dionysius' number system. However, by 1884 when an international conference in Washington DC selected the meridian passing through Greenwich observatory (just east of London) to be the marker from which 'Universal Time' or 'Greenwich Mean Time' was measured, the number zero was a standard part of mathematics, and so the Greenwich Meridian is considered to be 0 degrees of longitude, rather than 1.

Some important physical properties and the scales often used for their measurement are listed in Table 1.1. Note the way in which some of the choices for zero points are suggested naturally, others are more arbitrary.

1.5.3 Accuracy and exactness

The real numbers $\mathbb{R}$ give us a mathematical way of making sure, when measuring lengths or times, that two different states always have different numbers assigned to them. However, in practice we always make measurements of space and time to some greater or lesser degree of accuracy (as with the convention of giving only the year of most historical events). Whether it is centimetres, millimetres or microns, there is always a most fine-grained distinction, and within this interval,

---

[8]Hence the paradox of reckoning the birth of Jesus to have been roughly 4–6 years ahead of his time, according to most modern sources.

many states which are different in theory will be given the same measurement. When storing numbers and calculating with them on a computer, the same is normally true — the amount of memory (in binary digits or bits) which is allocated to storing a given variable puts an upper limit on the range of values which that variable can take, and the finest-grained distinction which that variable can hold. When rounding figures on a spreadsheet to the nearest whole number, or to three decimal places, you are really carrying out the same activity.

Generally, the degree of accuracy we choose to express is motivated by practical considerations, trying to be informative without splitting hairs. For example, when babies are born they grow and change very quickly, and to give meaningful comparisons we measure their age in months. Later in life, as there is less noticeable change from month to month and more months to keep track of than we care to count, we measure ages in years instead (Table 1.1).

For centuries, many scientists believed that this was a limitation imposed by our measuring instruments (or human diligence), but that the state of the universe itself was exact. Nowadays, and particularly due to the revolutionary results of quantum mechanics from the beginning of the 20th century, scientists realize that this is not really the case. Many measurements only make sense up to a particular degree of accuracy, and the more accurately we measure some variables, the more uncertain we are of others: for example, if we measure the position of a particle extremely accurately, this severely limits our ability to measure its movement.[9]

1.5.4 Continuous and Discrete quantities

The quantities we've discussed so far — time, length, angles — are all *continuous* quantities: however small an interval you consider, there are still infinitely many different possible lengths or angles within that interval. When measuring continuous quantities, the accuracy of our measurements is limited by the sensitivity of our instruments and the fine-grainedness of information that is appropriate to our descriptive purpose.

For some measurements, we do not have to choose a level of

---

[9] This is called the *Uncertainty Principle*, and was first formulated by Heisenberg (Bohm, 1951, Ch 5). Some of the ideas of quantum theory will have a big influence on the later chapters in this book.

granularity in this way, because the quantity being measured itself has a natural least unit of granularity. If I count the number of books on my desk at the moment (there are 3), or the number of keys on my computer keyboard (there are 105), this is using a number to describe a state-of-affairs in the world around, and so this also is a measurement. However, it is a measurement of a *discrete* rather than a *continuous* quantity.

The distinction between discrete and continuous quantities corresponds quite closely to the difference between *digital* and *analogue* signals. For many quantities, fractions just don't make sense, because the objects being measured come in whole-number chunks. Books, chickens, shoes and people are all examples. Sometimes discrete measurements take smaller-than-whole number amounts: for example, an amount of money might be measured in fractions of a dollar, but it's still discrete because the smallest fraction is one cent and you can't usually go any smaller than this, and if we counted in cents rather than dollars we'd have only whole numbers. One rule of thumb is that it often makes sense to ask "what would be the next biggest measurement?" with discrete quantities, but not with continuous quantities, because discrete measurements have a fixed smallest increment while continuous measurements do not.

Linguistic information comes to us in both continuous and discrete forms. A written text, once it is typed into a file on a computer, is a step-by-step list of characters, and the set of possible characters is itself discrete — each character is drawn from a finite set of letters, spaces, numbers and punctuation marks, each of which is represented by the computer using a whole number in some binary code.

On the other hand, when listening to speech, you are hearing sound waves which are continuous in nature, and in the process of interpreting these sound waves, you mark them off into particular words and phrases. In a sense, the speech becomes mentally segmented into a collection of discrete chunks of language. Another way to think of this is as a process of digitizing the speech signal — when a conversation or a speech is transcribed by a stenographer into a written text, they are effectively taking an analogue signal and producing a string of written characters, which is a digital or discrete format. The fact that many important things such as tone of voice and gestures may not be properly represented in the transcribed text, even though they may be a vital part of the original conversation or speech, is a warning that this model, like any other, is a simplification of reality.

All of the linguistic data we analyze in this book is drawn from textual sources which are kept as files in a computer (either transcribed speech or data that is naturally written text, such as newspaper articles), and so all of our measurements will be discrete. However, much of the mathematics we use to analyse these measurements will be based on continuous methods.

The classification of quantities into discrete and continuous groups has varied over the centuries — for example, Aristotle describes lines, surfaces and time as continuous (as we have done), but classifies both numbers and speech as being only discrete, saying that their parts have "no common boundary."[10] Today we think of some numbers, such as the real numbers $\mathbb{R}$, as continuous: but as we've seen, Aristotle lived in an age where faith in numbers as a description for continuous quantities (such as the length of a hypotenuse in a triangle) had been abruptly shattered, so this difference is readily understandable. However, it is certainly true that the positive, whole numbers (to which Aristotle is probably referring) are discrete in nature, and we shall turn to this set next.

1.5.5 The Natural Numbers $\mathbb{N}$

The most important set of numbers for measuring discrete quantities is the 'natural numbers', referred to by the symbol $\mathbb{N}$. These are the standard positive whole numbers with which you learnt to count, i.e. the set $\{1, 2, 3, 4, \ldots \text{etc.}\}$. Some mathematicians include zero in this set, but often zero is not regarded as a natural number. The natural numbers form the basis for all the other number systems, which can conceptually be thought of as extensions to the natural numbers which are needed in order to solve more and more problems.

For example, if we add two natural numbers a and b, the answer is another natural number $a + b$. However, if we subtract a from b, the resulting number $b - a$ will not be a natural number if a is greater than b. To make sure that $b - a$ is still a number, we extend the natural numbers

---

[10]*Categories*, Ch 6. Aristotle is referring to 'speech' once it has been segmented into discrete syllables, rather than the continuous sound waves we can nowadays measure directly in a phonetics laboratory, which I think accounts for this difference of description.

with the negative whole numbers (and zero), giving a set called the *integers*, which comprises *all* the whole numbers. The set of integers is given the symbol $\mathbb{Z}$ (from the German *Zahlen = numbers*). Similarly, if we multiply two integers a and b, the result ab will be another integer, but if we divide them the result $\frac{b}{a}$ might not be an integer (whole number) at all. For $\frac{b}{a}$ to be a number, we need to add fractions to the scheme, obtaining the *rational numbers* which are given the symbol $\mathbb{Q}$ (which could be thought of as "$\mathbb{Q}$ for Quotients"). We've seen that the fractions aren't enough if we want expressions such as $\sqrt{2}$ to have a number as their outcome, and we need to add irrational numbers to give the *real numbers* $\mathbb{R}$. But this still isn't enough to give a number for expressions such as $\sqrt{-1}$, because since 'a minus times a minus equals a plus', the number a^2 is *positive* for all real numbers (apart from $0^2 = 0$). To have numbers for quantities such as these we introduce the *imaginary* numbers (to distinguish them from the real numbers), starting with the symbol $i = \sqrt{-1}$, and the combination of real and imaginary numbers is called the *complex numbers* $\mathbb{C}$. Note that the conceptual jump here is *no different* from that required to add irrational numbers: in both cases we want to have a number for the expression $\sqrt{a}$, where a is an integer.

This gives a whole chain of sets of numbers,

$$\mathbb{N} \subset \mathbb{Z} \subset \mathbb{Q} \subset \mathbb{R} \subset \mathbb{C},$$

each set including the one before it. The whole structure is summarized in Table 1.2. This may seem like quite a daunting menagerie: the comforting part is that most of the time it will be obvious from the context what sort of numbers we're using and it is unlikely that the lack of a detailed knowledge of these mathematical structures will prevent anyone from understanding most of this book.

Strangely enough, the natural numbers were the last of these systems to be defined through formal axioms (the *Peano Axioms* for natural numbers were given by the Italian mathematician Giuseppe Peano in 1889, though they were also listed by Dedekind in the previous year), partly because it turns out to be quite difficult to formalize these basic counting objects, and partly because counting in whole numbers is so intuitive to us that it was not even seen as necessary to define them formally. Thus our story of mathematical formalism nears its end,

TABLE 1.2 Symbols for the different sets of numbers, and examples

Symbol	Kind of numbers	Examples
$\mathbb{N}$	Natural numbers	$1, 2, 3, \ldots, 4327, \ldots, 2^{467}, \ldots$
$\mathbb{Z}$	Integers	$-4327, \ldots, -2, -1, 0, 1, \ldots, 2^{467}, \ldots$
$\mathbb{Q}$	Rational numbers	$1, \frac{1}{2}, -\frac{17}{3}, 3\frac{1}{7}, \frac{8}{4}$
$\mathbb{R}$	Real numbers	$1, -\frac{4}{7}, \pi = 3.14159265\ldots, -\sqrt{2}$
$\mathbb{C}$	Complex numbers	$1, i = \sqrt{-1}, 2i = \sqrt{-4}, \pi - \sqrt{2}i$

with axioms for the whole numbers being given some 2200 years after Euclid's axioms of geometry.

1.6 Numbers in everyday language

Numbers are often regarded as the most absolute and rigourous of concepts — as an example of a statement that is incontrovertibly true, people will often quote that "2+2 = 4." However, the abstract concept of number comes long after the practical process of counting, and as we've seen, the formal properties of numbers were codified millenia after the formal properties of shapes.

The priority of counting over number is not only historical: it is clearly in evidence in the way we use number words. For example, in English, number words are much more often used as adjectives than as nouns. If you're at a market stall and ask for "nine," you will of course be questioned further as to "nine of *what*?" — the request for "nine" on its own is both grammatically and semantically lacking, and needs more information from the surrounding context for its meaning to be interpreted. If someone asks "how far?" or "how long?" something is and you answer "nine," you will likewise be asked whether you mean nine kilometres or nine miles, nine hours or nine days.[11]

---

[11]More substantial discussion of both the definition and use of numbers (and the ease with which a mismatch can occur between the *meaning* and the *application* of number words) can be found in Aristotle (*Metaphysics*, Book 10) and Frege (*Grundlagen der*

As we've seen, answering these questions depends on the conventions for using numbers to make measurements. The way real numbers are used to make continuous measurements depends on choosing a mapping between the system being measured and the real numbers, which almost always involves choosing a (possibly arbitrary) point of reference. Discrete sets of objects, on the other hand, often determine their own points of reference — if there are 3 books on my desk then 'measuring' this with the number 2 by starting to count from -1 rather than 0 would be absurd. The *total* number of objects in a given set is entirely determined by that set, and this property can be used to *define* whole numbers (called 'cardinal numbers') through set theory, independently of the Peano Axioms. However, there still needs to be some set of objects in mind for the statement "there are three" to make sense.

> **Number scarcity**
>
> Though in theory there are infinitely many natural numbers, in many systems we have been known to 'run out' of them, because in practice the length (number of digits) of a number is limited by necessity and by convention.
>
> Telephone numbers are usually expected to follow a regular format (for example, a 3-digit area code followed by a 7-digit local number), and when such a system reaches its limits, it must be altered by introducing new area codes or making numbers longer. Many companies increase the number of lines available through a single external telephone number by issuing extensions.
>
> Computing has introduced several constraints on numbers in different situations, ranging from limits on the length or accuracy of different data-types to the once notorious 'year 2000 problem.'

The result of this context-dependence is that, in practice, number words are among the most ambiguous in the whole language. The number nine is given as a name to literally millions of different objects — tables in different restaurants, mailboxes in different post offices, roads, junctions, streets and houses. This plasticity is licensed because once we have narrowed down our search to a particular context, we expect the number nine to have been used with the same unique distinction that it

---

Arithmetik), whose partly philosophical inquiries into the problem finally led to an abstract definition of numbers.

enjoys in the theoretical world. If you were going to a party and were told by a friend that it was "at the red house on Chester Crescent," if you got to the right street and there were two red houses, you would probably curse your friend's oversight but would accept it as a mistake. If you were told that the party was at "number 9, Chester Crescent," arrived at the street and found that there was more than one house with the number 9, you would be much more indignant at the incompetence of any city authority which could allow such a flagrant breach of the rules. The exactness of numbers *in a particular context* allows the same number to be used in many different contexts to mean many different things: a convention which, as we have seen, makes advances such as positional notation possible.

As well as uniqueness, there are other conventions in natural language which exploit the structure of numbers to give helpful names to objects. In a restaurant, if there is a single table right in between table 13 and table 15, we would guess that its number is 14. In this way, we expect the management to have exploited the *ordering* of natural numbers to give people a better chance of guessing the whole layout from a few examples. Even more so with street numbering: we expect the numbers to go up in one direction and down in the other, and there is also a very useful convention in many parts of the world which uses the odd numbers for one side of the street and the even numbers for the other.[12] Numbers are indeed a powerful tool for measuring and naming physical objects and situations, but their uses are *not* absolutely defined by *a priori* mathematical rules: the variety of different conventions, each tailor-made to ease a certain descriptive purpose, is a tribute to the plasticity of numbers and the ingenuity (not to mention opportunism) of human beings.

1.7 Conclusion

Many readers who have got this far may be pleased to find that the heading *conclusion* is being used to mean an ending, not a summing up of all that we have demonstrated. We've finished our tour of most of the fundamental mathematical concepts that will be needed to understand

---

[12]There is a famous anecdote of an ancient street in Germany where the houses were numbered *chronologically* as they were built: that this story is regarded as humourous is further evidence of how entrenched our conventions for using numbers to name objects has become.

this book — these are the building blocks with which our geometry of meaning will be described. If a lot of the material has been unfamiliar to you, then pat yourself on the back for covering so much new ground.

We've also seen that some of the building blocks of mathematics have been used for thousands of years without being formally described and understood. Life is too short and the universe is too interesting to worry about every exact detail: if you choose to press on anyway, you're following in the tradition of many great mathematicians.

Wider Reading

The most important single piece of wider reading I would suggest for this chapter, and as an introduction to this book, is to read at least some of Euclid's *Elements*, either online[13] or in one of several good versions in print, such as (Heath, 1956). There's no better introduction to what geometry is, and to the formal method of deducing results from a few axioms, which is how much of mathematics and logic works.

Other important primary sources can be found in Smith (1929): relevant among these works are Stevin's deliberately readable pamphlet which brought decimal fractions to the masses (1585, pp. 20–35), an extract from Descartes' analytic geometry which did more than any other work to enable the use of numbers to solve geometric problems (1637, pp. 397–402), and Dedekind's founding contribution to the theory of irrational numbers and the continuity of the real line (1872, pp. 35–45).

For readers who find set theory difficult, there are a variety of good introductions, in particular Chapter 1 of Partee et al. (1993). Jänich (1994), an undergraduate text on linear algebra which we will refer to quite often, contains an excellent introductory chapter which covers sets, maps, set operations, products of two sets and projections, all in a few pages: the reader who masters this material will find much of the rest of this book easy-going. For those interested in delving a little deeper into the mathematical side, Felix Hausdorff's (1914) book on set theory is really very readable and covers a great deal of ground.

For those interested in the different kinds of numbers and how they arise, I recommend Chapter 1 of Hamilton (1982). You'll find that, even though real numbers are pretty intuitive, the mathematical analysis necessary to prove that every real number can be expressed as an

---

[13] aleph0.clarku.edu/~djoyce/java/elements/toc.html

arbitrarily long decimal sequence, and that these sequences obey all the usual properties under addition and multiplication that we expect them to, is pretty fiendish: it was finally completed by Richard Dedekind (1831–1916), some 2400 years after the crisis in Greek number theory made the need for this work apparent.

On the more mystical subject of numerology, a notable example is that given in Buckminster Fuller's *Synergetics* (Fuller, 1975, 1200.00). The inventor of the geodesic dome entertained a lifetime fascination for geometry and numerology, and he wrote this special chapter as a poem.[14]

The correspondence between the natural numbers and the size of a given set was developed by Georg Cantor (who made the idea of 1-to-1 correspondence between sets precise in 1878), and by Gottlob Frege (1884) (who used this concept to define cardinal numbers). Cantor (1895) also realized that the concepts involved were by no means limited to finite quantities, and his work is instrumental in showing that the numbers we have defined (natural, rational and real numbers) are fundamental models for many discrete concepts (such as "the next number") and continuous concepts (such as "in between"). Frege's work includes a much more thorough analysis of the way numbers are *used* than is possible to summarize here: in developing a proper account he also laid the foundations for modern logic, as described online in the Stanford Encyclopedia of Philosophy (Zalta, 2003).[15]

Our discussion of mathematical approaches to the concept of space has been of necessity brief and simplified, and many mathematicians have been very interested in the nature of space itself. Ancient Greek and monotheistic ideas of space, the conflict between the 'absolute space' of Isaac Newton (1642–1727) and the 'system of relations' of Gottfried Liebniz (1646–1716), and subsequent developments up to nearly the end of the 20th century, are described by Jammer (1993).

The history of mathematics is a fascinating tale, and one that is rarely told. In almost every other discipline, as you learn about it you also learn about the writers and scholars who made the subject what it is. Not so in mathematics: students are usually just shown the bare bones

---

[14]This astounding work of modern science and historic insight is available on the web at www.rwgrayprojects.com/synergetics/synergetics.html.
[15]http://plato.stanford.edu.

of techniques that are used to solve certain problems, as if the techniques fell to us from heaven on tablets of stone. Little wonder that many are left with the impression that mathematics is an esoteric and difficult subject, with no human face. To dispel this myth, I strongly recommend a good history such as Boyer and Merzbach (1991): the first seven chapters discuss the prehistoric origins of mathematics, the ancient Egyptians and Mesopotamians (mainly the Babylonians), and Greek mathematics from the age of the Pythagoreans to the time of Plato and Aristotle, and finally Euclid himself.

2

The Graph Model: Networks of Concepts

In this chapter we describe the first and simplest of our models for the meanings of words, and in particular the meanings of nouns. By observing which pairs of nouns occur together in a large collection of texts, we will build and explore a network or *graph* where each noun is represented by a *node* with *links* to nouns which are similar in meaning. At least, that is the plan, though as we will see, the model throws out a few howlers as well as good links. In this way, our case study highlights some of the pitfalls — and the benefits — of building mathematical models directly from text data, and will hopefully raise some questions for the reader about how appropriate our model is for representing the meanings of words. As with any mathematical model for real-world phenomena, there is a trade-off between how exactly we want the model to describe the real-world situation, and how difficult the model should be to build — and to a large extent, the balance between these issues is mediated by the purpose for which the model is intended.

Before we describe the graph model itself, we will go through three preliminary stages. In Section 2.2, we describe some of the resources and tools for natural language processing that form a starting point for our investigations, and then in Section 2.3 we give a very brief introduction to what we mean by a graph. But before either of these, readers may find it well worthwhile to consider some of the issues behind defining words in terms of their similarities with other words.

2.1 Similarities between words

One might argue that describing a word in terms of other words to which it is connected is hopelessly circular (and of course, as has been pointed out many times, the same is true of a traditional dictionary). Most if not all of the models we talk about in this book will be open to this same criticism, so it is as well to give it some thought right now. When learning about the concepts in language directly from text, we are bound to learning about language in terms of what's in the text — which is other language. It's much harder to find out about the relationship between language and the world it's used to describe. For example, try defining the difference between the concepts *right* and *left* purely in terms of other words, without going round in circles. Not easy, is it? It's not until you start bringing in arbitrary conventions learnt from experience (such as "The side of the road people drive on in the USA, not the side people drive on in the UK") that you really feel you're making progress — but even then, this seems like a very poor definition of *left* and *right*, because at the end of the day this definition is circular as well.

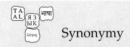

Synonymy

Two words that can be used to refer to the same concept, such as *violin* and *fiddle*, are called *synonyms* (the Greek roots being *syn* = *together* and *nym* = *name*). Very few pairs of words (if any) *always* mean the same thing in all contexts — it wouldn't normally be appropriate to substitute a violinist for a fiddler in a folk band, or a fiddler for a violinist in a symphony orchestra. Ever since Samuel Johnson's first dictionary (1755), lexicographers have warned against presuming that a synonym can be substituted for a particular word without taking the context into account. A list of synonyms can still be an invaluable aid to finding information in a standard directory of businesses and services (like the *Yellow Pages*), giving many useful pointers such as 'Car Stereos — see Automobile Stereos'

Is this circularity a problem chipping away at the foundation of our whole purpose? This already brings us back to comparing the study of objects with that of relationships. If we want to define once and for all what *things are* and what *words mean* then circularity would be a big problem. But from a geometric point of view, nothing could be more traditional than to study words by examining the relationships between them, and if this is circular then so is geometry.

From a different tradition, the arrangement of words into groups with similar meanings is the principle upon which Dr. Roget organized his groundbreaking new *thesaurus* in 1852, an insight which has given us one of the classic reference works of all time. Following this work, lexical resources which give lists of similar words are called *thesauri*.[1] As well as versions of Roget's general thesaurus, there are many thesauri available for particular fields or domains — for example, the Unified Medical Language System,[2] provided by the United States National Library of Medicine, combines several medical thesauri into a single 'metathesaurus' (Bodenreider and Green, 2001). Nowadays a thesaurus may contain all sorts of information over and above the traditional list of words and list of classes that each word is part of — thesauri are used as the backbone of many specialized information systems (such as MEDLINE), and many have several features designed to enable the indexing and retrieval of documents.

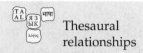

Thesaural relationships

One typical purpose of a thesaurus is to classify information so that a reader can find it in a systematic manner — for example, this is the goal of the Library of Congress Subject Headings (LCSH) and the Dewey Decimal System. Many modern thesauri contain standardized *equivalence, hierarchical* and *associative* relationships, often given by the codes UF (use for), BT or NT (broader term or narrower term), and RT (related to), which facilitate the search process. Our understanding and use of these relationships has increased greatly over the past 100 years, as the story of the LCSH bears witness (El-Hoshy, 2001).

The issues involved in building a thesaurus include choosing canonical forms for terms, arranging them into classes, coping with terms in multiple classes, using these terms to index documents which may be relevant to a variety of contexts, describing the kinds of

---

[1] The word *thesaurus* is of Greek origin and means a treasury or storehouse — so Roget's thesaurus is a treasury of knowledge. The plural *thesauri* is used overwhelmingly in computational linguistics, possibly because *thesauruses* sounds even more like we're talking about dinosaurs.

[2] www.nlm.nih.gov/research/umls/

relationships that arise between terms, finding the main features of concepts which distinguish them from other concepts, and maintaining a thesaurus when language is forever changing and growing. All of these are questions which natural language processing also faces, and many of the challenges we describe in this book have been considered by generations of librarians and lexicographers already.

Most of the geometric models we can build directly from corpora have much more in common with a thesaurus than with a traditional dictionary. With a dictionary, it is clear from the outset where a new term should be placed, determined by alphabetical order. But once this is accomplished, we are often no nearer to knowing anything about the meaning of the word — being told that *masquerade* comes after *masonry* and before *mass* in the dictionary doesn't tell us what it means at all.[3] On the other hand, a thesaurus is designed so that two words with similar meanings will occur in similar places — as Roget himself said in the introduction to his thesaurus,

> ...the words and phrases of the language are here classed, not according to their sound or their orthography, but strictly according to their signification.

It follows that a computer program which successfully finds groups of words with similar meanings would produce an output much more like a traditional thesaurus than a dictionary. Plenty of promising research has been done with precisely this goal in mind, to the extent that in the field of information retrieval *automatic thesaurus construction* is a recognized technique for enriching a search engine (Grefenstette, 1994), and the resulting data-structure is sometimes even called a *similarity thesaurus*. Some of the models later in this book can also be used for automatic thesaurus construction, so we will return to this theme more than once.

2.2 Some Tools and Terminology in Natural Language Processing

This section presents some of the techniques and approaches that have become standard fare in natural language processing, insofar as they

---

[3]Though words with the same root or stem are usually gathered together in a dictionary, and so *masquerade* is related in meaning to *masquerader* and *masquerading*, because they are all *derivations* from the same lexical root (Jurafsky and Martin, 2000, Ch 3).

have been used in this particular book. As in any field there is much that is taken for granted in the way things are usually presented at the conference and journal level (indeed, one of the more exciting things in natural language processing is that the level of technology which is taken for granted increases every year) and this section will hopefully explain some of these things so that the rest of the book does not leave behind those readers who are not already computational linguists. Those seeking a more thorough introduction to empirical natural language processing should consult a textbook such as Manning and Schütze (1999) or Jurafsky and Martin (2000).

Corpora

First of all, a large collection of texts used in natural language processing is called a *corpus* (for a 'body' of texts) and if you have many such collections they are called *corpora*. The most standard example of a *domain-general* corpus currently used by many linguists is the British National Corpus, or BNC. Other standard corpora include newspaper articles (the New York Times and the Wall Street Journal are among the most widely used) and official government records (which sometimes have the benefit of being translated into more than one language).

Corpora are collected for

The BNC

The British National Corpus (www.hcu.ox.ac.uk/BNC) was designed to reflect as broad a range of modern British English as possible. It contains around 90 million words from written texts, including samples of writing from newspapers, journals, published and unpublished letters, academic books, popular fiction, and much more. In addition, there are some 10 million words of transcribed speech from a variety of contexts, from radio broadcasts to business meetings, and everyday conversational samples collected by volunteers drawn from diverse demographic, social and regional groups.

research purposes by institutions such as universities and corporations, and in particular, many are made available by the Linguistic Data Consortium.[4] One benefit of using corpora is that researchers have access to language as it has been actually used, as opposed to how it

---

[4]www.ldc.upenn.edu/

theoretically *should* be used. Irregularities and exceptions all go into the melting pot, and this is vital if we're going to build systems that can eventually cope with the language they might encounter in a hitherto unseen document. So corpora are not in general sanitized or necessarily even good examples of their languages. Secondly, it usually takes considerable effort to get texts collected from a variety of sources into a common format where they can all be analyzed at once. If we're using data from the web, for example, it's important to strip out the human language from the computer language bits such as HTML markup. Although we don't want to interfere with the natural language itself, we certainly need to sanitize the formats so that we know which bits of the documents we're dealing with *are* natural language. Another important step might be removing or changing people's names or other sensitive information — especially when corpora deal with topics such as medicine or terrorism, the people who were involved in the events described in the corpus often should not be identified either deliberately or accidentally. Since this preparation can be a time-consuming process, it's usually much better for one research group or data consortium to collect a corpus and then to make it available to others, rather than for everyone to build their own. As well as the efficiency of this approach, as in many fields, there are many benefits to using data sets that are at least standardized, so that researchers can genuinely compare their results.

Ideally, a corpus will have a few desirable properties. The main one is simply size. The more language we have to train a system or model on, the richer that model will be, and the more information it will contain. However, adding new documents willy-nilly will not always improve a corpus because it may alter the *balance* of the corpus. Adding many versions of the same, or nearly the same document, will lead to redundancy and possibly skewed results. Also, not all corpora are designed to be as broad in coverage as the BNC — quite the opposite, in some cases. When designing an information system for a particular field of knowledge or *domain*, a corpus of documents from that domain can be very useful.

For example, suppose you want to create a dictionary of business terms. By examining the Wall Street Journal to see which terms do not appear in a standard dictionary, you can get a good first approximation of which terms will have to be added to make your business dictionary. For purposes such as these, it is important that your corpus accurately reflects the way language is used in that domain, for which *domain-specific corpora* such as the Wall Street Journal (Marcus et al., 1993) and the Ohsumed corpus of medical abstracts (Hersh et al., 1994) have been developed. On the other hand, to model language as broadly as possible, a corpus should have as many different kinds of language as possible, a goal which the BNC meets very well.

Types and tokens

When we say "there are around 100 million words in the BNC," does this mean there are 100 million different *kinds* of words or that there are 100 million words *altogether*? In this case it means the latter, though it is an ambiguous question. To resolve this ambiguity we distinguish between the number of *types* (distinct word forms) and the number of instances or *tokens* (total number of words) in a corpus. Perhaps best illustrated by an example, the quote

> To be or not to be ...

contains 6 tokens but only 4 different types: *be*, *not*, *or* and *to*. The number of tokens of a particular type is called the frequency of the type.

To a large extent, the notational conventions in this book are supposed to distinguish between types and tokens. If a word or phrase appears in *italic sans-serif* font, it is being used to identify a *type*, or the corresponding point or region in a geometrical model which represents that type. If a word or phrase is being used as a particular *token* or instance, double quotes are used to indicate that (for example) "this is an example usage."

The process of dividing a corpus up into distinct word-units is called *tokenization*, and is a nontrivial task even for English. (It's much harder in a language such as Chinese which doesn't place convenient whitespace between words.) Should hyphenated words be separated into distinct tokens or left as one token? Should a possessive form such as *John's* be treated as a type in its own right? This raises new questions about measurement — words in text may be pretty discrete objects, but clearly, the simple question "How many times does the word *John* appear in a particular corpus?" is not completely independent of measuring conventions and subjective choices. Can single concepts which are written as several words (such as *New York*) be treated as types in their own right? Even in our tiny example "To be or not to be," should *To* and *to* be considered as versions of the same type? This works for the two tokens in this example, but could cause a lot of confusion for the words *bush* and *Bush*. What we consider to be the number of types and tokens in a corpus clearly depends on how we choose to count, before we've even done any natural language processing. We won't delve into these questions very deeply, but introduce them here to give readers some idea of the baffling array of decisions that have to be made, every one of which could have an effect on the outcome of any experiment.

Part-of-speech tagging

A part-of-speech tag is a mark in a corpus telling us whether the corresponding word is a noun, verb, adjective, pronoun, etc. Part-of-speech tags can nowadays be added to corpora by pretty reliable computer programs called *part-of-speech taggers*. The British National Corpus, which we used for our experiments, is already tagged for parts of speech by the CLAWS2 grammatical tagger (Leech et al., 1994), which also segments the text into word and sentence units. As a result, instead of just having the input

> When Captain Pugwash retires from active piracy he is amazed and delighted to be offered a Huge Reward for what seems to be a simple task.[5]

we have the much more richly structured

> When&AVQ-CJS; Captain&NP0; Pugwash&NP0; retires&VVZ; from&PRP; active&AJ0; piracy&NN1; he&PNP; is&VBZ; amazed&AJ0-VVN; and&CJC; delighted&AJ0-VVN; to&TO0; be&VBI; offered&VVN; a&AT0; Huge&AJ0; Reward&NN1; for&PRP what&DTQ seems&VVZ to&TO0 be&VBI a&AT0 simple&AJ0 task&NN1.&PUN

which tells us (for example) that *Huge* is being used as an adjective, *Reward* is being used as a (singular) common noun, etc. In general, noun tags begin with an N, verb tags with a V and adjective and adverb tags with an A: the tags are explained in more detail in the BNC online documentation.[6] The hybrid tags (such as *amazed*&AJ0-VVN) occur when the tagger thinks both tags are possible: in this case there is a close call between tagging *amazed* as an adjective or a verb. These hybrids

---

[5]This example is from www.hcu.ox.ac.uk/BNC/what/garside_allc.html
[6]www.hcu.ox.ac.uk/BNC

are called *ambiguity tags* and occur with about 4.7% of the tokens in the corpus. According to the BNC documentation the tags have an error rate of only 1.7%, so we have a pretty reliable source of grammatical information at our disposal.

Contemporary taggers such as CLAWS2 tend to combine grammatical rules with statistical patterns. For example, the tagger will know that the word **Huge** is almost always an adjective but that the word **Reward** can be used as a verb as well as a noun: in the fragment **Huge Reward** it surmises that **Reward** is a noun because its preceding word is an adjective and so is much more likely in English to be followed by a noun than by a verb. In reality tagging is rarely this simple or sequential: see Manning and Schütze (1999, Ch. 9,10).

For our purposes in this chapter, the contribution of part-of-speech tagging is that it enables us to single out the nouns in the corpus (possibly preceded by modifying adjectives), and to see when they are linked by a conjunction such as *and* or *or*. This very slight amount of grammatical information proves to be enough to build an interesting graph of the way nouns are related to one another.

2.3 What is a graph?

This section is mainly for readers who aren't familiar with graph theory. The word **graph** may be encountered in many contexts — with a Greek root meaning 'write', **graph** has come to refer to many kinds of diagrams such as pie charts and lines connecting several points, and has given us the modern term **graphics**, which strangely enough now refers to the parts of a document which are *not* written material. *Graph theory* in mathematics is derived from the same theme, but in this context, to make the possibilities more manageable, we are more specific about what is meant by a graph. Readers who are already familiar with the basic notions of graph theory should be able to skip this section: for those who want more, there are some very readable introductions such as Bollobás (1998) and Chartrand (1985).

A graph is a collection or points or objects which are called *nodes*. In order to understand the relationships between these objects, certain pairs of nodes have *links* or *edges* joining them together. Thus a graph may be thought of as a very typical geometric space in the sense of Chapter 1: all we have is a set of points, and a list of relationships between pairs of points, and it is these relationships which give the space its structure. For example, in the graph in Figure 2.1, there is a link between node a and node c, and this link is given the name (a, c). This is similar to the way the vertices (corners) of a triangle might be labelled A, B and C, in which case the line in between A and B is called AB, and the nodes of a graph are sometimes also called its *vertices*. The formal properties of a graph are defined as follows:

Definition 1 A *graph* G consists of a set of *vertices* or *nodes* V and a set of unordered pairs $E \subseteq V \times V$ called *edges*.

The fact that $E \subseteq V \times V$ ("E is a subset of the Cartesian product $V \times V$") tells us that each element of E consists of a pair of elements of V — in other words, each edge is defined by its two endpoints. The fact that these are *unordered* pairs tells us that it doesn't matter whether we write $(a, b) \in E$ or $(b, a) \in E$: both tell us that there is an edge in between the nodes a and b, which can be travelled both 'from a to b' or 'from b to a'. As we shall see in Chapter 3, this need not be the case — in many situations, links in a structure only point in one direction. But for this chapter, the edges in the graph will all be two-way streets.

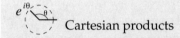 Cartesian products

Let M and N be two sets. An element of their *Cartesian product* $M \times N$ is a pair consisting of an element of M and an element of N: so if $m \in M$ and $n \in N$ their Cartesian product is the element $(m, n) \in M \times N$. Cartesian products are named after Descartes, who realized that any point in the plane can be represented by a pair of real numbers, its x and y coordinates, as we will see in chapter 5.

A daily example could arise in a restaurant where the set of meals on the menu is called M and the set of possible drinks is called D. In picking a meal m and a drink d, you would be choosing an element (m, d) from the Cartesian product $M \times D$.

Example 1 Consider the simple graph in Figure 2.1 (overleaf). It has four vertices a, b, c and d, so the set V is $\{a, b, c, d\}$. The edges E consists of the set $\{(a, b), (a, c), (b, c), (b, d)\}$, which is a subset of $V \times V$. The whole structure — both V and E together — is a graph G.

This collection of symbols may seem a much more fiddly way of writing things down than the diagram in Figure 2.1, and it's much easier to "see" the structure in the diagrammatic representation. The formal definition in terms of algebraic symbols has great power because

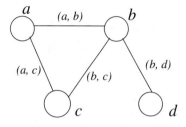

FIGURE 2.1 A simple graph.

it conveys a whole general class of objects: however, as soon as people talk about particular graphs they usually start drawing pictures.

If there is a link between two vertices v_1 and v_2, they are said to be *neighbours* of one another, in which case we will write $v_1 \leftrightarrow v_2$. So in our graph G, because (a, c) is an edge, we say that a and c are neighbours or just $a \leftrightarrow c$.

An enormous variety of situations can be described using graphs, which is one of the reasons that the concept is so useful. For example, a public transportation network can be thought of as a graph where each station is a node and to get to your destination node you traverse different edges in this graph.

FIGURE 2.2 The London tube.

Figure 2.2 shows a map of the London tube (underground railway) system which makes excellent practical use of this idea to help users to find their way without being confused by details such as the actual physical twists and turns of the different lines and the surface streets they pass under, which would not be relevant for the journey planning process. In fact, the tube map has more structure than just being a graph, because many of the individual edges between pairs of stations are collected together to form *lines*, such as the 'District Line' and the 'Circle Line' (if you'll pardon the paradox), and rather than just shuttling in between two stations, each individual train will normally run along a whole line. Sometimes a journey can be made just using one line, and otherwise you have to change trains, which takes longer. Allowing about 2 minutes to traverse each edge between stations and about 5 minutes each time you have to change trains, you can use the tube map to give a rough estimate of how long a journey in central London will take to complete.

We will talk about measuring distances in graphs and other geometric spaces in Chapter 4 — for now, note that the shortest distance between two stations in the tube system is not the traditional 'as the crow flies' straight line. In most human (and natural) situations, distances depend on the space you're traversing and what means are available to get around.

There are many other situations that can be thought of as graphs. Electrical circuits are an obvious example, and graph theory can be a very useful way of framing questions such as "How much current will flow between two components in my circuit?" (Bollobás, 1998, Ch 2). The distribution network for a manufacturing company can often be modelled as a graph,

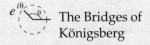 The Bridges of Königsberg

Graph theory often traces its roots to this famous problem. The city of Königsberg (modern day Kaliningrad) sits on the river Pregel which flows through its centre with two islands in its stream, the four land masses being connected by seven bridges. The problem: could you take a walk and cross each bridge once and once only?

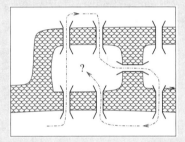

When Leonard Euler (1707-1783) visited Königsberg, he was told about the problem by many citizens who had puzzled over it. By thinking of each land mass as a node in a graph and each bridge as a link, Euler not only proved that the problem couldn't be solved: he also worked out what properties *any* graph would need for such a circuit to be possible, namely that at most two of the nodes could have an *odd* number of links (Bollobás, 1998, p. 17). According to legend, Euler managed all of this before dinner was over.

where some of the nodes are places which produce goods (factories), some are places which consume goods (shops and stores), and some are in-between nodes where goods are collected along the way (warehouses). In this situation, each link (possible transportation route) between two nodes is usually given a numerical *weight* representing the cost in time or physical resources of traversing that particular link, enabling us to ask "what is the most economical path between two nodes?"

Example 2 A thesaurus can easily be turned into a graph by treating each word as a node, and saying that two words are linked if they are listed as synonyms or related words. For example, in the Mirriam-Webster on-line thesaurus, the words *setback, reversal, bill* and *tab* are listed as synonyms of (different senses of) the noun *check*. So we would build a graph G by giving *check* a node (written formally as $check \in V$) and placing a link between *check* and *setback* (written formally as $(check, setback) \in E$), and so on for all the other nouns and links between them.

In conclusion, the main themes of this section are that

- A graph is a pretty simple object, consisting of a collection of *nodes* and *links*.
- Graphs are a useful and very general way of describing many situations.

The next step is to see how we might use the idea of a graph to describe relations between the meanings of different words.

2.4 Building a graph of words from corpus data

In this section we describe how a graph was built to represent the relationships between the nouns in the British National Corpus. Since the corpus is already tagged for parts of speech, we knew to a high degree of accuracy which tokens were nouns. But how do you know just from this whether two nouns are similar in meaning?

The main intuition behind our answer is that when people write down a list of nouns, those nouns are usually related in meaning to one another. There are literally millions of very simple examples of this in standard webpages. If you've ever booked tickets for anything on-line, how many times have you seen a drop-down menu allowing you to choose between different months? If you also want to sign up to earn frequent flyer miles, the website will ask for your personal details, including which country you live in. A typical example (from the United Airlines website) is given in Figure 2.3.

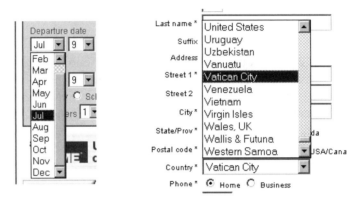

FIGURE 2.3 Months and countries from a webpage menu

Of course, the computer itself may not 'know' that these words refer to similar things — all it needs to do for the moment is to print out a string of letters. But with a bit of intelligence, the fact that these words occur together can be very useful. For example, suppose you hadn't heard of *Wallis and Futuna* before you saw the menu in Figure 2.3. Then based on the other more familiar terms in the list, you'd be in a good position to guess that *Wallis and Futuna* is a country, or at least a geographical region, and it turns out that you'd be right: Wallis and Futuna is a group of islands in the Pacific, and a French overseas territory.

Natural language does not, of course, come with the convenient formatting that tells a browser to put words in a menu together, and from which we can therefore infer that this list contains words which are related in some way. But there are other ways in which language presents us with lists of objects from which we can infer the same information. For example, consider the sentence

Ships laden with **nutmeg**, **cinnamon**, **cloves** or **coriander** once battled the Seven **Seas** to bring **home** their precious **cargo**. (BNC)

Thanks to the part-of-speech tags in the BNC, we know that the words in boldface are nouns. After looking at a lot of sentences like this, we started to guess that nouns separated only by commas, or the word *and*, were often related to one another, so that in the sentence above, **nutmeg, cinnamon, cloves** and **coriander** were good candidates to be linked to one another. More generally, for any nouns n_1, n_2 and n_3, the patterns

$$n_1, n_2, \text{ and } n_3 \qquad n_1, n_2, \text{ or } n_3 \qquad (2.1)$$

can be taken as evidence suggesting that $n_1 \leftrightarrow n_2$ (n_1 is a neighbour of n_2), $n_2 \leftrightarrow n_3$, and also $n_1 \leftrightarrow n_3$. We will refer to such patterns as *coordinations*, and say that (for example) the nouns n_1 and n_2 occur 'in coordination' (Huddleston and Pullum, 2002, Ch 15).

Several researchers have used simple templates of this sort to extract information. Just for the fairly specialized goal of automatic word learning (or 'automatic lexical acquisition'), extracting this kind of information from coordination patterns was pioneered by both Riloff and Shepherd (1997) and

> **Coordination one-liner**
>
> Perl programmers can get a first approximation to the coordination algorithm with one line of code: searching through the saved e-mail on my Stanford account with the one-line Perl script
>
> ```
> perl -e 'if(/John and (\w+)/)
>     {print "$1 "};' -n /Mail/*
> ```
>
> gives the output
>
> ```
> Yoko Melanie I ChiSook I
>     Roger Maryl I his I
> ```
>
> and if I part-of-speech tagged my e-mail and removed the pronouns from this output, I'd be left with a small collection of first names.

Roark and Charniak (1998). Before this, Hearst (1992) pioneered the technique of finding phrases matching a few special templates or *lexicosyntactic patterns* to extract semantic relationships. Part of the appeal is that it's a simple and intuitive idea, which also makes it easier to write into a program than most algorithms in natural language processing.

What this ends up giving us is a graph G, in precisely the sense of Definition 1. The set of nodes V consists of the nouns in the corpus, so each noun is a node in the graph. The edge set E consists of all the pairs (n_1, n_2) where the nouns n_1 and n_2 are observed in a coordination pattern like the ones in (2.1).

When we ran this on the whole of the British National Corpus, the resulting graph in the experiments described has 99,454 nodes (nouns)

56 / GEOMETRY AND MEANING

and 587,475 edges or links. This means that on average, each noun in the graph is linked to roughly 12 other nouns. There were roughly 400,000 different types tagged as nouns in the corpus, so the graph model represents about one quarter of these nouns, including most of the more common ones.

Many of the more telling links were observed more than once — for example, the *apple* and *orange* were observed in coordination no fewer than 18 times. Does this mean that the edge (*apple, orange*) should be regarded as more important than the edges (*apple, egg*), (*apple, adobe*) or (*apple, piece*), each of which were obtained from only one coordination pattern? As with transportation models, should we attach a 'cost' or 'weight' to each edge? The general question of *weighting* (how much importance to give to different measurements) crops up in many parts of natural language processing — for example, a search engine will have to decide how much weight to give to the observation that a particular keyword occurred in a particular document (Baeza-Yates and Ribiero-Neto, 1999, p 25).

> Programming for the graph model
>
> The graph was implemented using the following steps. Firstly, the pattern-matching technique for extracting the noun-noun relationships can be done using *regular expressions*, one of the simplest structures in formal language theory (Jurafsky and Martin, 2000, Ch 2) and easily programmed using Perl. The extracted relationships are then stored in a database (so our graph is 'really' a database in much the same way that a triangle is 'really' 3 marks on a piece of paper).
>
> The pictures of the graph in this book were produced using freely available graph visualization tools developed by AT&T and available through www.graphviz.org.
>
> You can explore the graph online at infomap.stanford.edu/graphs

The challenge of finding the best link-weighting strategy for our graph of nouns is still very much alive — we've tried various options, some of which are discussed more thoroughly in Chapter 4, but there isn't a clear winner yet because ideas that seem to work for some words might not work so well for others. For example, we could simply discount edges that are observed only once, because in many cases this seems to be just the sort of statistical noise that can arise in large corpora of natural language, where after all almost anything can happen. But many rare words have only one link and it's often a good one. For example, the only node that the *aristophanes* node is linked to is the *thucidides* node, which in turn is linked to *herodotus, aristotle* and *sophocles*, from which we would be (correctly) tempted to infer that Aristophanes was an Ancient Greek and probably an Athenian.

One well-behaved option was to take the highest ranking k neighbours of each word, where k could be determined by the user. In other words, if the system was asked 'give me 6 neighbours of the word *apple*' it would simply return the 6 neighbours of *apple* whose coordination with *apple* was encountered most frequently (and all the neighbours of *apple* if it had fewer than 6).[7] In this way the question of what weight to give to each edge was resolved simply by ranking the edges. One consequence of this decision was that edges linking to more common words were preferred over edges linking to rarer words. We suspect that this may have made the output of the graph more stable but less exhaustive (in technical evaluation terms, we may have boosted *precision* at the expense of *recall*). Finding the best way to assign weights to links in the graph for different tasks is still very much an open challenge.

2.5 Exploring the graph

Once we've got our computer to go through the corpus, build the graph, and store it in a database, what then? We could evaluate the model to see how well the neighbours and clusters we obtain from the graph correspond with neighbours we'd get from a thesaurus or from human judgments. If these evaluations show that the system is sufficiently reliable, we could even use the graph to predict useful information about the meanings of words which are missing from a dictionary or knowledge base. Experiments of this kind will be described in Section 4.5 — in the meantime, many readers will find it interesting, not to mention good fun, to take a short guided tour of the graph and see a few pictures of what it can accomplish.

---

[7]One side-effect of this method is that the graph may no longer be 'symmetric'. For example, *bach* is one of the top 3 neighbours of *telemann* (a less famous Baroque composer), but the top 3 neighbours of *bach*, naturally enough, are *handel, mozart* and *beethoven*, so *telemann* is pushed off the list. More on this in Chapter 4.

58 / GEOMETRY AND MEANING

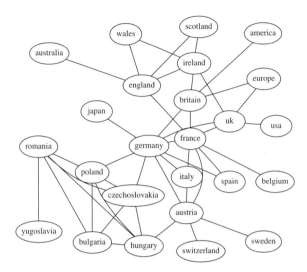

FIGURE 2.4 Growing a graph of countries.

Finding related countries

The diagram in Figure 2.4 was produced by starting with the *poland* node in our graph and clicking on related nodes, which added the 5 words with the strongest links to the target node, gradually building up a network. As you can see, most of the related nodes reached in this fashion are other countries, and the edges link these countries to related countries — often countries that share a geographical border, or otherwise a strong economic, cultural or historical link.

While it does very well at linking similar countries, the graph does a poor job of linking countries which instead of being merely similar are more or less synonymous (or one is part of the other). For example, the graph does not realize that *usa* and *america* or *britain* and *uk* are (often) used interchangeably, or that *england, scotland* and *wales* are part of *britain*. This isn't surprising, since phrases like "America and the USA" or "England and Britain" don't occur very often. It appears that our graph is better at finding related words than actual synonyms: if two words have the same meaning then using them in the same sentence would often be redundant, so they may not appear together very often in corpora.

As well as exploring the graph by hand, we can also start with an

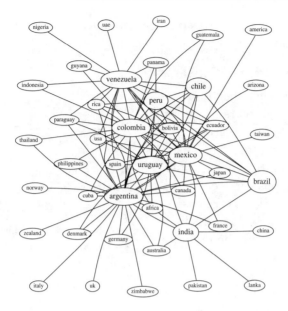

FIGURE 2.5 Words related to *brazil* in the graph.

initial or seed node and ask the system to give us its neighbours, the neighbours of its neighbours, etc. This is how the graph in Figure 2.5 was made, starting with *brazil*. This time we asked for the top 8 neighbours of *brazil*, giving 8 primary neighbours (the words in the larger ovals). The 5 top neighbours of each of these were then added, giving the secondary neighbours (the words in the smaller ovals).

Some of the single words that occur are only fragments of the names of countries, such as *zealand* and *rica* instead of *new zealand* and *costa rica*. This is because the model tries to get the most information possible out of its input data by stripping off modifying adjectives, so that for example

apples and green pears

will still give us a link between *apples* and *pears*. In many cases, this helps us to squeeze a lot more good information out of the corpus. However, for phrases like "Australia and New Zealand" this doesn't work too well, and we should really have preprocessed our data to recognize *New Zealand* as a separate 'named entity' before we started.

Still, the main thing is that the computer is doing a surprisingly good

2.6 Symmetric relationships

On the whole, the links in our graph stay within a pretty uniform level of abstraction: that is to say, countries are linked to other countries, not to cities, towns, regions or continents. There are some exceptions — possibly because this is built automatically from the BNC, we see a fair bit of "Britain and Europe," which likens an island to a continent. But on the whole, the links are quite horizontal.

One of the reasons we would expect this to be the case is that being joined together by a conjunction such as *and* or *or* is often a reversible operation: if you encounter the phrase "football and cricket," it's reasonably possible that you could also encounter the phrase "cricket and football."[8] Relationships which are reversible in this fashion are given a special name.

Symmetric spatial words

Some of the simplest words for describing the spatial relationships and comparisons between objects are symmetric or very clearly asymmetric. For example, if "*a* is *beside* *b*" it would normally be the case that "*b* is *beside* *a*," so *beside* signifies a symmetric relationship. However, if "*a* is *below* *b*" then it *can't* normally be true that "*b* is *below* *a*," so *below* is not a symmetric relationship (and as we shall see shortly, *below* is actually *anti*symmetric).

Definition 2 Let $\sim$ be any relationship between mathematical objects. Then $\sim$ is said to be *symmetric* if the statement $a \sim b$ always implies that $b \sim a$, and vice versa.

Probably the simplest exam- ple of a symmetric relationship in mathematics is the familiar 'equals sign' for numerical quantities: if $a = b$ then $b = a$ and vice versa. Many day-to-day relationships are also symmetric: if A is a sibling of B then B is also a sibling of A, and similarly for cousins and spouses.

---

[8]Though there are many more or less idiomatic preferences to this construction: for example, "cat and mouse" sounds a lot more natural than "mouse and cat" (Huddleston and Pullum, 2002, p. 1287)

Example 3 Symmetric relationships in graphs

In the graphs we've used in this chapter, 'being a neighbour' is a symmetric relationship. This was ensured in Definition 1 by saying that the relationships in the graph are *unordered*, so that $(a, b) \in E$ and $(b, a) \in E$ are the same statement. This is precisely the same as saying that the neighbour relationship '$\leftrightarrow$' is symmetric, because if $a \leftrightarrow b$ then also $b \leftrightarrow a$. In practice, whenever we encountered the phrase

cat and mouse

in the BNC, the database entry for *mouse* was updated to link *mouse* to *cat*, and the entry for *cat* was updated to link *cat* to *mouse*.

It's also possible to have graphs where the edges are *not* reversible in this fashion, but instead are one-way streets, as will be described in Chapter 3. This can happen in several of the real-life examples of Section 2.3 — for example, if a train breaks down in between stations a and b on an underground railway and blocks the line in one direction, then the edge between the stations becomes a one-way or *directed* edge.

2.7 Idioms and Ambiguous Words

This section describes some of the characteristic shapes that occur in our graph around idioms and ambiguous words. The geometric properties of ambiguous words in the graph model will be analyzed more carefully at the end of Chapter 4.

Idioms

In British English, the idiom "chalk and cheese" is used to highlight a pair which are naturally *dis*similar, such as two people with very different tastes, as in

> They are psychologically, spiritually and in personality as different as chalk and cheese. (BNC)

or two institutions which operate very differently, as in

> The finances of the two museums are chalk and cheese, with taxpayers providing a mere 17% of the Met's annual budget but over 80% of the British Museum's. (BNC)

In our graph (Figure 2.6), *cheese* isn't linked to any rocks other than *chalk*, and *chalk* isn't linked to any foodstuffs other than *cheese*, leaving the link between *cheese* and *chalk* the only one holding between these unlikely areas of meaning like an axle holding together separate wheels.

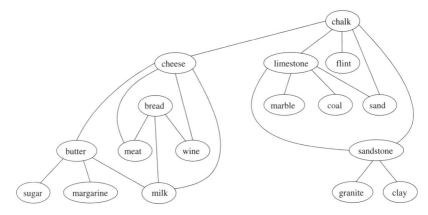

FIGURE 2.6 The 'pair of scales' shape characteristic of a particularly idiomatic phrase or link.

It's even possible for several idioms to lead us round in an unlikely circle. In Figure 2.7, *cat* and *mouse* are linked (not surprisingly, since they're both animals and the phrase "cat and mouse" is used quite often): but then *mouse* and *keyboard* are also linked because they are both objects used in computing. A *keyboard*, as well as being a typewriter or computer keyboard, is also used to mean (part of) a musical instrument such as an organ or piano, and *keyboard* is linked to *violin*. A *violin* and a *fiddle* are the same instrument (as often happens with synonyms, they don't appear together often but have many neighbours in common). The unlikely circle is completed (it turns out) because of the phrase from the nursery rhyme

> Hey diddle diddle,
> The cat and the fiddle,
> The cow jumped over the moon;

In spite of appearing to be a circle of related concepts, many of the nouns in Figure 2.7 are not similar at all, and many of the links in this graph are derived from very very different contexts.

Ambiguous words

Just as a special idiom such as "chalk and cheese" can become represented by a link in the graph connecting otherwise distinct regions, an ambiguous word can become represented by a *node* which has links to otherwise distinct regions. For example, Figure 2.8 shows that the

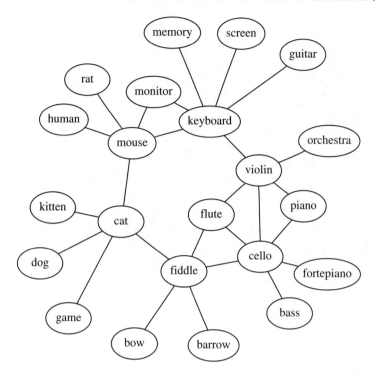

FIGURE 2.7 *Cat, fiddle, violin, keyboard* and *mouse* take us on a wild goose chase.

neighbours of the word *arms* fall into two categories, parts of the body being on the right and weapons on the left.

A similar division into regions corresponding to different senses is shown in the graph of the word *head*, parts of the body being on the left and words related to leadership on the right. In this graph, we've restricted our attention to *cycles* (loops) because this often seems to focus the graph within clearly recognizable areas of meaning.

In Chapter 4 we will have a closer look at ambiguous words, and in particular the way they affect the way we try to find shortest paths and measure distances between different concepts.

Conclusion

A lot of interesting information about how words are related to one another has been extracted simply by looking at which groups of words appear in lists together in everyday natural language texts, by doing

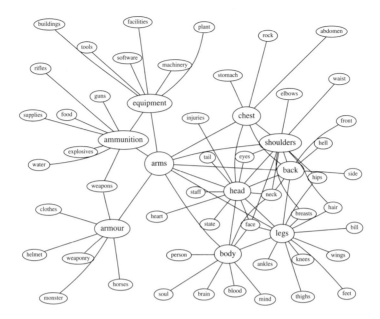

FIGURE 2.8 The neighbours of *arms* fall into two categories.

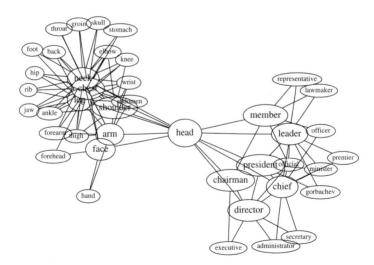

FIGURE 2.9 The neighbours of *head* fall into two categories, showing a 'bowtie' shape that is sometimes a sign of ambiguity.

some simple grammatical preprocessing and looking for patterns in the output.

This method can be used to build a whole graph of concepts, in which we can see interesting *geometric* patterns which are symptomatic of recognized *semantic* phenomena such as idioms and ambiguity.

Wider Reading

As well as Bollobás (1998), good (and refreshingly inexpensive) introductions to graph theory are to be found in Chartrand (1985) and Trudeau (1994), which combine an introduction to the pure mathematical concepts with several easily-grasped practical problems as motivation. Chartrand (1985) also contains a good introduction to weighted graphs or networks.

The circularity of dictionary definitions is a known problem, not just in theory but in practice. Many lexicographers have for years tried to follow the maxim of defining each word in terms of more simple, basic words: a famous example is the Longman Dictionary of Contemporary English (LDOCE), designed particularly for those learning the language, in which the definitions are given in terms of the *Longman Defining Vocabulary* of only 2000 common words (Proctor, 1978). This approach is conceptually not unlike the mathematical idea of deducing results from a collection of basic axioms, knowledge of which must be assumed, and leads to the questions "what should we consider to be the most basic semantic units?" and "do they differ or are they the same for different languages?." Those interested in such questions might begin with the Natural Semantic Metalanguage of Wierzbicka (1996), which discusses the *semantic primitives* into which different concepts may be factorized. Such theories are criticized by Aitchison (2002, Ch 7), on the grounds that there is no overall consensus on what these primitives should be, and that there is no evidence as yet that we use 'primitive' concepts any more quickly than 'compound' concepts, which would be the case if concepts in the mind were decomposed on-the-fly.

More about corpora and part-of-speech tagging can be found in the relevant sections from Manning and Schütze (1999) and Jurafsky and Martin (2000). In addition, our treatment of 'words' could be improved by recognizing groups such as **New Zealand** as a single entity. Such a group of words can be described as a *collocation* or more specifically a *named entity*: techniques for collocation extraction are discussed in Manning and Schütze (1999, Ch. 5).

A good introduction to many current thesauri and the issues in building them is given in Bean and Green (2001), and some of the papers in this collection have already been referred to. Foskett (1997) and Kilgarriff (1993) are challenging but readable articles concerning issues behind the creation and use of thesauri and dictionaries respectively; a broad overview of the role of lexicons in natural language processing is given by Guthrie et al. (1996).

There are many techniques for finding synonyms or related terms from text, only some of which will be described in this book. Miller and Charles (1991) is a classic paper which discusses some of the philosophical and psychological issues behind synonymy and similarity of meaning, testing the idea that words occurring in similar contexts are similar in meaning. Automatic thesaurus generation for information retrieval is discussed in several works (Salton and McGill, 1983, p. 78), (Grefenstette, 1994) and (Baeza-Yates and Ribiero-Neto, 1999, §5.4), while Lin (1998a), Pereira et al. (1993) and Schütze (1998) use similarity of distribution across a corpus for a variety of tasks such as identifying different senses of ambiguous words — in general, the level of accuracy and (apparent) intelligence with which a computer can discover which words have closely related meanings is astonishing to many layfolk.[9]

Many of the issues involved in adapting and using thesauri for natural language processing were first addressed in the 1964 PhD thesis of Karen Sparck Jones, which presents the theoretical issues behind synonymy and other semantic relations, and the learning of synonyms and classifications for purposes such as thesaurus construction and disambiguation. That this work was published (with an introductory chapter concerning the intervening years' progress) over 20 years after its completion Sparck Jones (1986) is a tribute to its lasting contribution to natural language processing.

The specific method of using coordination patterns to find similar words was pioneered by Riloff and Shepherd (1997) and Roark and Charniak (1998), using (in general) smaller corpora and more detailed

---

[9]Online examples can be found through Google labs.google.com/sets, Dekang Lin's personal website www.cs.ualberta.ca/~lindek/demos.htm (an excellent set of examples which also allows you to get clusters representing different senses) and of course infomap.stanford.edu/webdemo and infomap.stanford.edu/graphs.

linguistic analysis (including full parsing) than was used in this chapter. The idea of the graph model was first presented in Widdows and Dorow (2002): in this work, we devoted less attention to the grammatical processing used in building the model, and much more attention to the *combinatoric* (graph-theoretic) analysis of the model once it was built, yielding exceptional results which will be presented in Chapter 4. The extent to which linguistic sophistication, mathematical care and sheer corpus size contribute to the quality of results in this area is still a very open question.

3

The Vertical Direction: Concept Hierarchies

The previous chapter demonstrated a technique whereby a computer, with a very little knowledge of the way English works, can tell us that *hungary, poland, romania, bulgaria* and *czechoslovakia* are all related. So far so good, but this begs the question, *what are they*? The next step would be to have a technique for working out that they are all European countries. Something like Figure 3.1:

List of words (nouns): hungary czechoslovakia poland romania bulgaria Send

Class Label	Score
European country, European nation	3.500
Balkan country, Balkan nation, Balkans, Balkan state	1.250
country, state, land, nation	0.972
administrative district, administrative division, territorial division	0.458
district, territory	0.268
region	0.176
location	0.124
object, physical object	0.092
entity, something	0.072

FIGURE 3.1 Recognizing a group of European countries.

Or if we want to know what *cheese* and its top five neighbours *butter, milk, meat, bread, wine* from the left hand cluster in Figure 2.6 have in common, we would like to be able to take these words and classify them as in in Figure 3.2.

List of words (nouns): cheese butter milk meat bread wine   [Send]

Class Label	Score
foodstuff, food product	2.500
dairy product	2.250
food, nutrient	0.944
substance, matter	0.472
object, physical object	0.334
beverage, drink, drinkable, potable	0.250
entity, something	0.248
money	-0.250
combatant, battler, belligerent, fighter, scrapper	-0.250
baked good, baked goods	-0.250
dark red	-0.250

FIGURE 3.2 These neighbours of *cheese* suggest that it's a foodstuff and more specifically a dairy product.

Needless to say, there is an algorithm behind this class labelling trick, and it relies on a lot of careful work by many Princeton students over many years. The goal of this chapter is to describe this work and the mathematical ideas behind it.[1] Two of the most important characters in this story — Aristotle and Darwin — are not usually thought of as mathematicians at all, but nonetheless they described their ideas using mathematical models which were, if anything, far ahead of their time.

The idea is that concepts can be arranged into a *hierarchy*, a tree of meaning whose trunk and branches correspond to *general* concepts and whose twigs and leaves correspond to particular or *specific* concepts. We shall see that there are many examples of this sort of structure in common use, from the famous 'Tree of Life' to postal addresses and computer file systems. If the symmetric relationships of the previous chapter can be thought of as level or horizontal in character, the relationship between a child-node and a parent-node in a hierarchy (most clearly exemplified in the relationship between a species and its genus in the Tree of Life) can be thought of as a vertical relationship.

---

[1] The algorithm itself is running at infomap.stanford.edu/classes and you're welcome to try it out for yourself.

3.1 Phylogeny, or the Tree of Life

Aristotle (384-322 BC) probably contributed more to Western Science than any other individual before or since, and even though the content of many scientific theories has moved on since, a brief glance at some of his works should be enough to convince most readers that for better or worse, the style of academic writing is to this day thoroughly Aristotelean. He was born in 384 BC, in Stagira on the northern shore of the Aegean sea, and at the age of 17 he joined the brilliant group of men who had gathered in Plato's Academy in Athens. Aristotle worked and studied in the Academy for 20 years, but when Plato himself died in 347 BC, his former pupil, forsook a brilliant career in cosmopolitan Athens for a 12 year field trip of discovery.

The story of Charles Darwin and his voyage to the Galapagos Islands on the *Beagle* has become one of the great legends of modern scientific discovery. The story of Aristotle's years of fieldwork on the islands in the Aegean sea is less well-known, but curiously similar and no less remarkable. Both men left their accustomed careers for the opportunity to devote a few quiet years (and the utmost care) to observing the the habits, parts, structure, organs and reproduction of living things. Both described their findings in terms of species, genera and inheritance — individual offspring inherit the properties of their parents, and species inherit the properties of their corresponding genera. By collecting individuals into species and species into genera, the famous *phylogeny* or *Tree of Life* emerges (Figure 3.3).

The ideas behind this tree structure are developed in Aristotle's writings, and drawings of such trees derived from Aristotle's works have been extant at least since medieval times (as shown in the picture on page 80). Hundreds of years later, Darwin supported the tree analogy for evolution enthusiastically and explicitly, writing in *The Origin of Species*,

> The affinities of all the beings of the same class have sometimes been represented by a great tree. I believe this simile largely speaks the truth. The green and budding twigs may represent existing species; and those produced during former years may represent the long succession of extinct species. ... As buds give rise by growth to fresh buds, and these, if vigorous, branch out and overtop on all sides many a feebler branch, so by generation I believe it has been with the great Tree of Life, which fills

THE VERTICAL DIRECTION: CONCEPT HIERARCHIES / 71

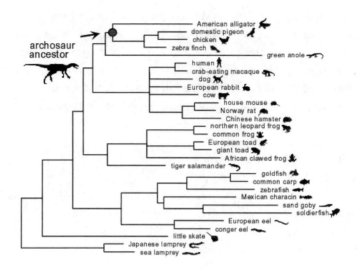

FIGURE 3.3 A Tree of Life, showing species alive today and the predicted *archosaur* (Chang et al., 2002, by permission of Oxford University Press).

with its dead and broken branches the crust of the earth, and covers the surface with its ever-branching and beautiful ramifications.

(Darwin, 1859, p. 172)

Aristotle's and Darwin's views agree on the overall validity of this structure, though Darwin's thought has largely superseded Aristotle's in arguing that the tree of life is the result of eons of evolutionary change. This advance owed a great deal to other scientific knowledge which was unavailable in Aristotle's day: in particular, the science of geology gave Darwin an enormous timespan to work with, onto which different stages of his evolutionary tree could be matched using the record of fossils.

Aristotle wrote down many of his findings in his works on *History of Animals* and *Parts of Animals*. Much of the language we use today when describing the animal kingdom is directly descended from Aristotle's works, in particular, the idea that each individual belongs to a particular *species* and each species belongs to a broader *genus*. The idea that some pieces of information are more *specific* and others are more *general* has been borrowed from these roots and is used to describe knowledge in every modern field. Aristotle himself uses the idea in his logical works: if a property is known to be true of a whole genus, we can argue that it is also true of each species (and each individual) belonging to that genus. (For example, if we know that a *horse* is a *mammal* and that all mammals have four limbs, we can deduce that all horses have four limbs.) This is a typical example of the way properties are *inherited* from genus to species.

The structure of the tree of life is one of the core examples of a *taxonomic* approach to classification, and any such tree structure will often be called a *taxonomy*. We will come to think of this structure as a collection of nodes and links, a bit like the graph model in Chapter 2. Each of the final leaves and each of the junctions in the tree of life can be thought of as a node (the leaves are called *terminal nodes*, the junctions are sometimes called *internal nodes*). The links in the graph (branches in the tree) might connect an individual to its species, a species to its genus, and so on.

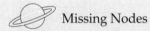 Missing Nodes

Darwin argued that the ancestry of all living things could be traced through the tree back to a primitive root node representing the first, most basic living things. There are many nodes in this tree which have no living representatives, which appears to be a problem for Darwin's theory: however, Darwin argued that many of the nodes had *once* represented living species, but that these species were long since extinct (for example, the *archosaur* node in Figure 3.3).

In particular, Darwin's account suggested that mankind and our nearest relatives in the animal kingdom should have a common ancestor, which was commonly referred to as the *missing link*.

3.2 Directed Relationships

One of the main differences between the graph we use to represent a hierarchy such as the Tree of Life and the graph in Chapter 2 is that links in a hierarchy should *never* be though of as symmetric. If we know that a every *horse* is a kind of *mammal* (so that there is a link from the *horse* node to the *mammal* node), we would not assume that a *mammal* is also kind of *horse* (which would correspond to placing a link in the opposite direction).

There are many many pairs of words in English whose relationships are asymmetric in nature, as in the Doormouse's famous musing from *Alice in Wonderland*,

> 'You might just as well say,' added the Doormouse, who seemed to be talking in his sleep, 'that "I breathe when I sleep" is the same thing as "I sleep when I breathe"!' (Carroll, 1865, Ch 7)

which is a joke precisely because *sleep* and *breathe* shouldn't be swapped around like this, because their relationship is not symmetric but asymmetric.

Many other relationships in both mathematics and language are *not* symmetric. For example, for two numbers a and b, if $a < b$ (a is less than b) then it won't be true that $b < a$, because a can't be both greater and less than b. Among family relationships, if a is a parent of b then b can't be a parent of a. Unlike symmetric relationships, the *direction* of these relationships is critical to interpreting what information is being conveyed. In English syntax, word order is often used to make the direction of the relationship clear: for example, the statement

"Malvolio loves Olivia"

asserts that there is a directed relationship between *Malvolio* and *Olivia*, but unfortunately for Malvolio it doesn't necessarily follow that

"Olivia loves Malvolio."

Many words that describe directed relationships have a counterpart which describes their opposite or converse: for example, if a is *below* b then b is *above* a. Sometimes a pair of very similar words is used to describe the different directions of a relationship, such as *employer* and *employee*.

Since there are clearly many directed relationships to be considered, we adapt some of our mathematical definitions to cope with them. Fortunately this is no sooner said than done, by changing the unordered pairs in the definition of a graph (Definition 1) to ordered pairs.

Definition 3 Directed graphs

A *directed graph* G consists of a set of *nodes* or *vertices* V and a set of *ordered* pairs $E \subseteq V \times V$ called *directed links* or *directed edges*. If $(a, b) \in E$ we write $a \rightarrow b$.

In the same way that we use the notation $a \leftrightarrow b$ for a typical symmetric relationship, we will use the notation $a \to b$ for a typical directed relationship (though this will not appear very often, because we will soon focus on special directed relationship such as those between a species and its genus, and these relationships will be given a more specific symbol). Just as there are many examples of graphs, there are also many examples of directed graphs.

3.2.1 The World Wide Web and *PageRank* algorithms

A good example of a directed graph is the World Wide Web, where each webpage can be thought of as a node and each hyperlink between pages as a link in this graph. The insight that the World Wide Web can also be modelled as a directed graph is largely responsible for the *PageRank* algorithm (Brin and Page, 1998) which made the Google search engine[2] so successful.

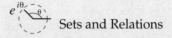

Sets and Relations

Definition 3 could also be presented by saying that E is a *relation* on $V \times V$, which is just a subset of $V \times V$. Thus a (directed or undirected) graph is a very standard object in set theory and mathematical linguistics (Partee et al., 1993, §2.2). In fact, directed graphs are in a sense more general than undirected graphs, because for a directed graph E may be *any* relation on $V \times V$, whereas for an undirected graph, E must be a *symmetric* relation. In spite of this, the standard approach in graph theory is always to presume that a graph is undirected by default, and to introduce directed graphs as a special class.

The *PageRank* algorithm treats the web as a directed graph, where each webpage a is a node and $a \to b$ if the page a contains a hyperlink to the page b. (So you can click on a link in page a which takes you to page b, or a contains a *citation* of b.) The intuition is that if many different webpages have included a link to page b, then many authors have found the page b to be useful or interesting, so it is more likely that the user of a search engine will also find the page b interesting. Thus the simplest 'pagerank' score for b would be obtained by counting the number of different pages a which contain a link to b, i.e. the number of nodes a such that $a \to b$, or more formally $|\{a : a \to b\}|$, where for any set A, $|A|$ is the number of elements in A.

---

[2] www.google.com

Google's *PageRank* is more sophisticated than this in at least two respects. Firstly, a page which contains hundreds of links is casting its vote much more thinly than a page containing only one or two links, so the contribution to the *PageRank* score made by each link from the page a is divided by the total number of links made by a. Secondly, links from pages which themselves have a high ranking are accorded more significance: if the American Mathematical Society were kind enough to include a link to my homepage on their website, this would be a much greater accolade than if my dad decided to include a link to my homepage from his. Thus the *PageRank* of a page b, PR(b), is determined by

> **Recursion**
>
> Sometimes an algorithm or process uses the output of one stage as the input for the next — for example, the *PageRank* algorithm of Equation 3.2 uses the rank of a page a as part of the calculation of the rank of each page b to which a has a link. Many mathematical and computing techniques involve some form of recursion, as do some family relationships. For example, the *ancestors* of a include the parents of a, the parents of the parents of a, the parents of the parents of the parents of a, and so on.

$$\mathrm{PR}(b) = (1 - D) + D \sum_{\{a : a \to b\}} \frac{\mathrm{PR}(a)}{C(a)}, \qquad (3.2)$$

where the sum Σ is taken over all pages a that link to b and $C(a)$ is the number of citations (outward links) contained in the page a, so $C(a) = |\{c : a \to c\}|$. The constant D between 0 and 1 is a damping factor, so that not all of a page's rank has to be determined by the number of citations it gets. (Brin and Page, 1998, §2.1.1)

3.3 Antisymmetric relationships and trees

It's perfectly possible that in a directed graph, $a \to b$ and $b \to a$ may both be true: for example, two webpages may easily be linked to *each other*.[3] However, for many directed relationships, this inversion cannot happen: for many interpretations of the directed relationship $\to$, if $a \to b$ then it *can't* also be true that $b \to a$.

---

[3] Such two-way links on the web are called *co-links* by Eckmann and Moses (2002), who demonstrate that they are significantly more useful than unreciprocated one-way links as a step in uncovering thematic layers (groups of pages about a common theme) on the web.

We have already come across some such relationships. If a is above b then b cannot be above a, and if a is a parent (or ancestor) of b then b can't be a parent (or ancestor) of a. The same holds for some mathematical relationships between numbers. If a is greater than b (written $a > b$) then it *can't* also be true that $b > a$. If we loosen this condition to $a \geq b$ ("a is greater than or equal to b"), then it is also possible that $b \geq a$, but only if a and b are equal. (For example, if $a \leq 6$ *and* $a \geq 6$, it follows that $a = 6$.) Relationships of this sort are the opposite of symmetric relationships.

Definition 4 Let $\sim$ be any relationship. Then $\sim$ is said to be *antisymmetric* if whenever the relationship $a \sim b$ holds, the relationship $b \sim a$ *cannot* hold unless a and b are identical.

An important family of antisymmetric relationships is given by the notions of containment and implication introduced in Section 1.2.1. If a set A is contained in a set B (written $A \subseteq B$) then we cannot also have $B \subseteq A$ unless A and B are the same set. In the same way, if the statement A implies the statement B, and the statement B also implies the statement A, then it follows that the two statements are equivalent.

Hierarchical relationships, such as those between a species and its genus, are also antisymmetric — as stated already, if we know that a *horse* is a kind of *mammal*, it can't also be true that a *mammal* is a kind of *horse* (unless the classes of *horses* and *mammals* are exactly the same, which they aren't). A hierarchy or tree can actually be defined as a collection of nodes connected by antisymmetric links in such a way that a unique chain of relationships can be traced from each object to the root of the tree (Davey and Priestley, 1990, p. 26). (The image might be even better if we called this unique ancestor node the *trunk* of the tree, since the roots of a tree spread out just as much as the branches — however, I hope the *root* metaphor is intuitively clear.)

Throughout this book we will use the notation $a \sqsubseteq b$ for a hierarchical relationship, which could be interpreted as meaning (every) a is a kind of b or the set objects described by a is contained in the set of objects described by a. Before we examine the uses of these relationships in linguistics, we will consider a couple of examples of the way information is often described in hierarchical terms: that is, as trees of concepts linked by antisymmetric $\sqsubseteq$ relationships.

THE VERTICAL DIRECTION: CONCEPT HIERARCHIES / 77

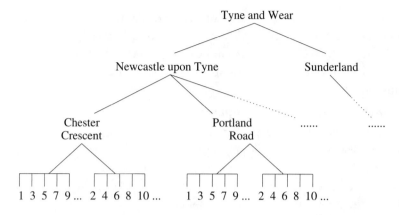

FIGURE 3.4 Part of the address tree of a region in the UK

Postal addresses

Consider the address

9 Chester Crescent, Newcastle upon Tyne, Tyne and Wear, UK.

This postal address gives a chain from the leaves at the bottom (individual houses) to the root node at the top, as shown in Figure 3.4. Reading along, we start at the smallest conceptual unit, the number *9* which tells us where in the street *Chester Crescent* the house is. *Chester Crescent* refers to a street in *Newcastle upon Tyne*, the biggest city in the region of *Tyne and Wear*, which in turn is a part of the *UK*. The address is a chain of directed relationships such as *Newcastle upon Tyne*

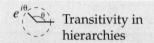 Transitivity in hierarchies

In general, if $a \sqsubseteq b$ and $b \sqsubseteq c$, it will also be the case that $a \sqsubseteq c$. For example, from *Chester Crescent* $\sqsubseteq$ *Newcastle upon Tyne* and *Newcastle upon Tyne* $\sqsubseteq$ *UK*, it follows that *Chester Crescent* $\sqsubseteq$ *UK*. Relationships of this sort, where from $a \sim b$ and $b \sim c$ we can infer that $a \sim c$, are called *transitive*, a property we will study in Chapter 4. Hierarchical relationships such as hyponymy are (normally) transitive, as Aristotle points out in Chapter 3 of the *Categories*.

$\sqsubseteq$ *UK*. To deliver the letter, the postal service works in the opposite direction, starting at the top of the hierarchy and working down.

As we read along the address (up the tree), the names refer to more general concepts, and *vice versa* — towards the left (down the tree), the concepts are more specific. So how far up or down the tree we are can be used as a measure of generality or specificity, and specificity in this sense is one of the well-known semantic dimensions (Cruse, 2000, p. 50). If there is a relationship $a \sqsubseteq b$, it will usually be the case that a is more specific than b.

Directories (or folders) on a computer

All of us who use computers will be familiar with directories or folders, of the sort pictured in Figure 3.5. Here again we have a tree structure, where each folder is contained in the folder above it. Such hierarchies are fairly ubiquitous throughout computing: for example, an XML document (a bit like the HTML a browser reads but more general) has a natural tree-structure, and the objects in object-oriented programming languages (such as Java and C++) are in a hierarchy, so that a new object can inherit the properties (such as variables and methods) of the object above it.

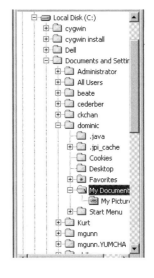

FIGURE 3.5 Directories or Folders

This property of inheritance is one of the main reasons why trees are such a useful way of compressing information. If we know that a *car* has *wheels* and that a *BMW* is a kind of *car*, we can infer that a *BMW* also has *wheels*. In the address above, once we know that *Newcastle upon Tyne* is in the north of England, we know that this particular *Chester Crescent* is also in the north of England. Using a conceptual structure which naturally enables inheritance can be very useful for describing groups of species and genera in biology, and for inference in logic, and these benefits motivated both Darwin and Aristotle's use of tree structures.

Our examples show that there are many ways of representing trees and hierarchical information, and humans are adept at developing conventions for exploiting this structure without even thinking about it. Just from the fact that something is written on the front of an envelope, you *know* how the names of the towns and streets are related even if you haven't heard of them before. Computer scientists and linguists know that the convention in their fields is that when someone draws a tree, the root is placed at the top (as in Figure 3.4) or sometimes at the left (as in Figures 3.3 and 3.5 — but hardly ever at the bottom! Part of the great challenge for natural language processing is to mimic even a small portion of this human awareness of which situations have which conventions for representing the same information.

Note that most of the names in our trees are at least potentially ambiguous until they're seen in context. As described in Chapter 1, number symbols are particularly dependent on context — the 9 in "9 Chester Crescent" is the number given to millions of homes, but as long as there are no two houses with the number 9 on the same street, we're fine. With computers, you can give two different files the same name provided you don't try and put them in the same directory (in which case you may end up overwriting one of them). In many computer operating systems, *every* folder or directory has a folder with the simple name ".." which refers to the directory above it in the tree — so what the name ".." refers to depends entirely on which folder you're in.

 Relational nouns

Many simple relational nouns can be used to refer to many different people or concepts: for example, to know which person is being referred to by the word *mother*, you need to know whose mother it is. Effectively, a word like *mother* takes as input a node in a family tree, and returns the female node one generation higher up. The meaning of words that you might not naturally regard as ambiguous may still be completely dependent on the context in which they're used, and that this is second-nature for humans is one of the things that makes it so hard to implement for computers.

In this way, we see a primitive notion of *context* beginning to assert itself. Which node in a hierarchy a particular term refers to (for example, which house is meant by the number 9) is determined by the terms it occurs with. With the phrase "9 Chester Crescent," this task is made easy for us because **Chester Crescent** is also a node in the hierarchy directly above the 9 in question. If we skipped this stage and just said "House number 9 in Newcastle upon Tyne," finding the meaning of the ambiguous "number 9" would be much more difficult.

80 / GEOMETRY AND MEANING

 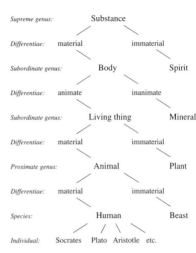

FIGURE 3.6 The *Tree of Porphyry*, one of the earliest examples of a concept hierarchy. From the ceiling fresco at Schussenried, Germany, by Franz Georg Herrmann (1757), photograph by Jeffrey Garrett (Northwestern University Library, October 2000).

3.4 Representing linguistic meaning in trees

In this section, we describe some of the ways that tree structures have been adapted to describe the relationships between concepts in language. The ideas are often traced back to Aristotle, a famous example being the *Tree of Porphyry* (Sowa, 2000), which gives a pictorial representation of some of Aristotle's *Categories* (Figure 3.6).

Porphyry's 4th century classification was based on dichotomies (yes/no questions) — is something substantial or insubstantial, animate or inanimate, and so on. In machine learning, such a structure is called a *Decision Tree* (Mitchell, 1997, Ch 3): a more familiar version of this way of classifying concepts is in the word game *Twenty Questions*, where the players have to try and guess the name of some concept by asking 20 yes/no questions of their choice.

The terms *genus* and *differentiae* in this diagram are traditionally used in lexicography to refer to the core classification of a concept and its more individual properties which distinguish it from other members of the same genus — this principle (and these terms) again going back to Aristotle's *Categories*. Dictionary definitions are often written down in such a form: for example, Merriam-Webster's definition of a (European) *robin* is

> 1) (a) a small chiefly European thrush (Erithacus rubecula) resembling a warbler and having a brownish olive back and orangish face and breast.

This tells us that *robin* ⊑ *thrush* (its genus), and that the differentiae of *robin* include resembling a warbler, having an orangish breast, *etc*. Darwin explains how the differentiae of species arises using the principles of *variation* and *natural selection*.

Many such links can be gathered into a whole knowledge structure, such as the *semantic networks* or *knowledge bases* that are used in many artificial intelligence systems (Lehmann, 1992, Sowa, 2000). Semantic networks can be very rich and varied, with many different kinds of links and relationships between concepts. One distinction that is often made is to talk about the hierarchical ⊑ relationships as taxonomic, and this part of the semantic network as a taxonomy. A more general

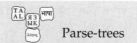

Parse-trees

Another crucial use of tree-structures in linguistics is to describe the grammatical relationships between words and phrases in a sentence. For example, the parse tree below tells us that the sentence (S) "The dog barked" is made up of a noun phrase (NP) "The dog" and a verb phrase (VP) "barked" (Jurafsky and Martin, 2000, Ch 9).

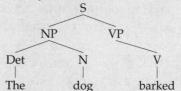

Parsing is more rewarding (and more difficult) than just part-of-speech tagging, since it involves finding not just part-of-speech for each word, but how these units relate to one another in a sentence — in old-fashioned terms, finding 'Who did what to whom?'

semantic network may include PART_OF or USED_FOR links, and such a knowledge base is often called an *ontology*. For example, UMLS, which is a medical ontology, has relationships such as TREATS which may link a drug or therapeutic process to the condition it is supposed to help, and CAUSES which may link a kind of bacteria to a disease, or a drug to known side-effects.

Needless to say, along with the growth of ontologies, the terminology used to describe them has also proliferated. Aristotle describes the relationship $x \sqsubseteq y$ by saying that y is *predicated of* x. The same term is used throughout his writings on logic: if y is predicated of x you can infer that anything that is true of y is also true of x, so in using this term for the relationship, Aristotle is also teaching his students how to *reason* with it, an important application of modern knowledge bases.

In linguistics, the name given to the hierarchical relationship $\sqsubseteq$ is *hyponymy* from the Greek roots $hypo = beneath$ and (once again) $nym = name$ (Cruse, 2000, Ch 8). The concept a is a hyponym of the concept b if (every) a is a kind of b. For example, *apple* is a hypernym of *fruit* and *greyhound* is a hyponym of *dog*, which is itself a hyponym of *mammal* — and of course, *greyhound* is also a hyponym of *mammal*, because every greyhound is a mammal as well as being a dog. The converse of a hyponym is a *hypernym* ($hyper = above$, which is confusingly similar to its opposite, $hypo$). If a is a hyponym of b (so that $a \sqsubseteq b$), we say that b *subsumes* a.

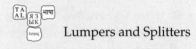

Lumpers and Splitters

How many senses for a particular word should be recorded in a dictionary or knowledge base? We might expect the answer to be clear in most cases, but in practice there is a great deal of variation between different lexicographers and dictionaries. For example, should *board* meaning a surface on which to play a game be distinguished from *board* meaning a flat object used for a specific purpose (such as an *ironing board*), or is the former really a variant of the latter?

Similar questions arise in biology: Darwin notes that some of the most distinguished naturalists differ greatly on whether they count a group of creatures as a separate species or merely a particular variety, and nowadays these debates continue unabated over the internet.

Knowledge bases containing many taxonomic relationships can be very useful, but also very time-consuming and expensive to construct, which raises the challenge of learning these relationships automatically. Ideally this could be done from ordinary text, just as the relationships in the graph in Chapter 2 were learned from ordinary text. To some extent, the techniques we used to extract the symmetric relationships used in our graph can be slightly altered and used to extract hierarchical relationships. For example, by adding the word *other* to give the phrase "*a* and other *b*" can make all the difference between "France and Spain" (a symmetric or horizontal relationship) and "France and other European Countries" (a hierarchical or vertical relationship). The technique of extracting hierarchical relationships using patterns of this kind was pioneered by Hearst (1992) (see also Hearst (1998)).

One of the biggest differences between the Tree of Life and any linguistic taxonomy is the problem of ambiguity. The classification of living creatures into species and genera is accepted as being a pretty thorough and moreover *unambiguous* way of indexing known living creatures. However, no such exhaustive approach works as convincingly for natural everyday language, because word meanings evolve in a much freer fashion. For example, the word *mouse* has in recent years leapt across the tree of meaning from being a rodent

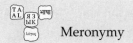

Meronymy

Another important relationship between some pairs of nouns is the *meronymy* or PART_OF relationships (Greek *meros* = *part*), so for example *steering wheel* is a meronym of *car* and *tail* is a meronym of *wolf*. The converse of a meronym is a *holonym* (Greek *holos* = *whole*), so *wolf* is a holonym of *tail*. Meronymic and hypernymic relationships are often interrelated: for example *France* ⊑ *European Country* and *France* PART_OF *Europe* are interdependent, and both are related to the idea of containment between sets.

to also meaning a small piece of computing equipment (Figure 4.12) — not because these meanings have any physical genetic material in common, but because a computer mouse is also small, usually grey, and has a tail.

3.5 WordNet

The taxonomy we will refer to most often in this book is the taxonomy of nouns contained in the *WordNet* knowledge base (Fellbaum, 1998b). The WordNet project was begun in the late 1980s by a group of researchers in linguistics and psychology at Princeton University, with the goal of modelling the way concepts might be stored in the mind. In the process, thanks to the diligence of many many researchers and graduate students at Princeton, WordNet has become the most widely used English-language lexicographic resource in computational linguistics at the present time, and the multilingual EuroWordNet was based on its design (Vossen, 1998).

WordNet gives information for words in four main categories, which are nouns, verbs, adjectives and adverbs. The modifiers (adjectives and adverbs) are organized into pairs of *antonyms* ('opposite names'), such as (*good, bad*) and (*heavy, light*) — most adjectives in English have an antonym of this sort which is a structure that most nouns and verbs lack (Miller, 1998b).[4]

The nouns in WordNet are organized hierarchically, so that the hypernyms (ancestors) and hyponyms (descendants) of a noun can easily be looked up (Miller, 1998a). For example, the possible hypernyms given for the word *oak* in WordNet 2.0 are

oak $\sqsubseteq$ wood $\sqsubseteq$ plant material $\sqsubseteq$ material, stuff $\sqsubseteq$ substance, matter $\sqsubseteq$ entity

oak, oak tree $\sqsubseteq$ tree $\sqsubseteq$ woody plant, ligneous plant $\sqsubseteq$ vascular plant, tracheophyte $\sqsubseteq$ plant, flora, plant life $\sqsubseteq$ organism, being living thing, animate thing $\sqsubseteq$ object, physical object $\sqsubseteq$ entity

A more pictorial representation of the hypernym trees above different European countries is given in Figure 3.8, and a few of the nodes from the very top of the WordNet noun hierarchy are shown in Figure 3.7: if the reader has any trouble picturing the way we are classifying *oak* and its hypernyms, these diagrams should help.

Note in 3.5 the way in which the ambiguity of *oak* is described in the WordNet taxonomy — the word *oak* has different chains of ancestors depending on whether we mean oak trees or oak wood. To some extent, this ambiguity is resolved by adding the term *oak tree* to one of the meanings of *oak*, effectively splitting the *oak* concept into two nodes, one just *oak* and the other *oak, oak tree*. This list of terms which can all be used for the same concept (such as *life form, organism, being, living thing*) is called a *synset* in WordNet, the idea being that the *concept*

---

[4]Aristotle points out this difference in properties: things in the category of *relatives* usually have contraries, and the examples he gives are mainly adjectives. On the other hand, *substances* "have no contrary," and these are almost all nouns.

represented by a node in the hierarchy is the conjunction or intersection of all the individual *terms* in the synset. In this way, WordNet attempts to give relationships between concepts rather than just relationships between the words which represent them.[5]

Of course, *oak* ⊑ *living thing* is just as true as *oak* ⊑ *tree*. However, *oak* is certainly *nearer* to *tree* than it is to *living thing*, and since there are no nodes in between *oak* and *tree*, we can say that *tree immediately subsumes oak*, or that there is just one step from *oak* to *tree*. In this way, we begin to define a primitive notion of *distance* in a taxonomy — the number of links in between two nodes, where each link must be *as short as possible*, in the sense that it can't be broken down into a chain of smaller links.

Using WordNet

WordNet is freely available from www.cogsci.princeton.edu/~wn/ which includes documentation, related papers, instructions for installing WordNet on your own system, and an online browser if you want to look up individual words and find their relations.

At the time of writing, WordNet 2.0 has recently become available, which includes more topical organization and new links between nouns and verbs which are morphologically derived from the same root.

A few of the top nodes in the WordNet hierarchy are shown in Figure 3.7, which gives the reader a sense of the main categories of meaning into which nouns are classified.[6] These nodes are the ones deemed important enough to have a whole file devoted exclusively to them in the WordNet source files. Some of their names are technical (or have a technical meaning that is being used in this particular context): for example, the term *artifact* is used to refer to

---

[5]To a considerable extent, the multi-faceted nature of words and their meanings contributes to making WordNet more complicated than just a simple tree, where each node has a unique name and a unique path can be traced from each node to the root of the tree. Since the hyponymy relationships are directed, the existence of two upward paths from *oak* doesn't enable us to go round in circles — nonetheless, most texts on graph theory would not strictly speaking call the WordNet noun taxonomy a tree. The ambiguity and context-dependence which is so natural to the structure of word senses is a significant challenge in adapting traditional mathematical models to describe meaning in natural language. Using *synsets* instead of just words in WordNet is a very good attempt to address this difficulty.

[6]An interesting (and quite fun) exercise in *ontology alignment* is to try and work out the correspondence between the top of the WordNet noun hierarchy in Figure 3.7 and the ten categories discussed by Aristotle, which are translated as *substance, quantity, relation, place, time, position, state, action* and *affection*.

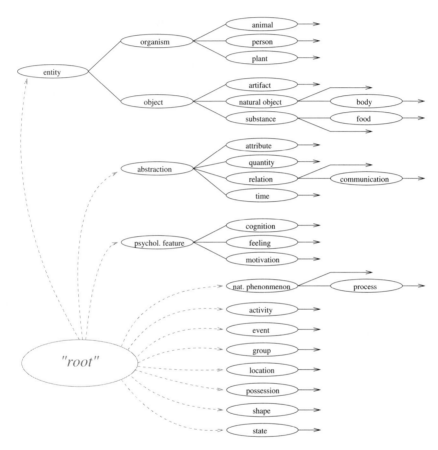

FIGURE 3.7 A few of the topmost nodes in the WordNet noun hierarchy (Miller, 1998a, p. 30)

any man-made object, so a *sewing-machine* is just as much an artifact as (for example) a *bronze-age amulet*, which would probably correspond more closely to what the word *artifact* conjures up in most of our minds. Other terms such as *state* are highly ambiguous words and are being used with their most general senses.

The eleven orphan nodes at the left hand side of Figure 3.7 are called *unique beginners* in WordNet: these are not subsumed by any more general parent nodes in WordNet and are therefore the roots of their respective trees. In practice, we often want to think of the nouns in WordNet as a single tree (for example, for working out the distance between two nodes along a shortest path). To do this, we may introduce an over-arching *root node* which subsumes *every* other node.

Not all of the possible concepts in language have words attached to them — and indeed, the choice of *which* concepts should have words attached to them differs from language to language. For example, both *fingers* and *toes* are collectively referred to as *dedos* in Spanish (sometimes *dedos del pie* is used to refer specifically to *toes*). To avoid leaving gaps, WordNet has several entries for concepts which are not expressed by any one English word, but which are significant enough to justify a separate node in the hierarchy: examples include *wheeled vehicle*, *living thing* and *body part*.

The verbs in WordNet are also organized in a hierarchy, but things turn out to be more complicated with verbs — "they've a temper," as Humpty Dumpty points out in *Through the Looking Glass* (Carroll, 1872, Ch 6). As well as what sort of action they refer to, verbs are also affected by the number and nature of the participants in the action (their 'argument structure') and how they can be made up of several events with different temporal relationships (their 'aspect'). For example, it is true that "I breathe when I sleep," and that *sleeping* logically entails *breathing*, but it would seem very strange to say that *sleeping* is a kind of *breathing*. However, it does make sense to say that *ambling* is a kind of *walking*. WordNet distinguishes these relationships as MANNER_OF relationships, calling this relationship *troponymy* (Fellbaum, 1998a, p. 79).

Due to these complications, we must admit that the geometric structures in this book make much clearer sense for representing the meanings of nouns and nominal expressions generally, while a representation of verbs as points in some space remains conceptually difficult, partly because many important properties of verbs depend on the way they combine with their arguments (subjects, direct objects, indirect objects, etc; Huddleston and Pullum 2002, Ch 4), and to model this in a geometric space we need a geometric method for combining points into more complex units. Aristotle makes little attempt to categorize verbs, giving (almost as an afterthought) six sorts of *movement* and various senses of the term *to have* (*Categories*, Ch. 14, 15). Verbs are discussed in Aristotle's treatise *On Interpretation*, but only as *parts* of propositions, their "proper meaning" being alluded to but never analyzed.

3.6 Finding class labels: Mapping collections of words into WordNet

As promised at the beginning of this chapter, we will now explain how the class labels in Figures 3.1 and 3.2 were obtained, giving the reader a chance to look at the workings. In a nutshell, the algorithm that provides these class labels tries to find the concept which has the most relevant vertical relationships with the list of input words, using a lexical knowledge base called WordNet to provide these vertical relationships. In other words, the highest scoring class label is the concept which subsumes (is a hypernym of) as many of the input words as possible, as closely as possible.

The algorithm works as follows. Let the word w be a noun (or a verb) in WordNet and let $\mathrm{Hyper}(w)$ be the hypernyms (or troponyms, for verbs) of w as given to us by WordNet.

$$\mathrm{Hyper}(w) = \{h \text{ such that } w \sqsubseteq h\}.$$

We can extend this definition to give the hypernyms for a whole set S of input words, defining $\mathcal{H}$ to be the union

$$\mathcal{H} = \bigcup_{w \in S} \mathrm{Hyper}(w). \qquad (3.3)$$

Our intuition is that the most appropriate class label for the set S is the hypernym $h \in \mathcal{H}$ which subsumes as many as possible of the members of S as closely as possible in the hierarchy. There is a trade-off here between subsuming as many as possible of the members of S, and subsuming them as closely as possible. This line of reasoning can be used to define a whole collection of class labelling algorithms.

For each $w \in S$ and for each $h \in \mathcal{H}$, define the *affinity score function* $\alpha(w, h)$ between w and h to be

$$\alpha(w, h) = \begin{cases} f(\text{dist}(w, h)) & \text{if } w \sqsubseteq h \\ -g(w, h) & \text{if } w \not\sqsubseteq h, \end{cases} \quad (3.4)$$

where $\text{dist}(w, h)$ is a measure of the distance between w and h, f is some positive, monotonically decreasing function,[7] and g is some positive (possibly constant) function. For now, set the distance function $\text{dist}(w, h)$ to be the number of minimal $\sqsubseteq$ relationships (single links that can't be decomposed into smaller links) in the chain between w and h. For example, with the hypernyms given for *oak* (on page 84),

$$\text{dist}(\textit{oak}, \textit{tree}) = 1 \quad \text{and} \quad \text{dist}(\textit{oak}, \textit{living thing}) = 5.$$

This gives us a way of acknowledging that, although both *oak* $\sqsubseteq$ *tree* and *oak* $\sqsubseteq$ *living thing* are true, *oak* $\sqsubseteq$ *living thing* is a much bigger leap than *oak* $\sqsubseteq$ *tree*.

The function f in equation (3.4) accords positive points to h if h subsumes w, and the condition that f be monotonically decreasing ensures that h gets more positive points the closer it is to w. On the other hand, the function g subtracts penalty points if h does not subsume w. The function g could depend in many ways on w and h — for example, there could be a smaller penalty if h is a very specific concept than if h is a very general concept.

Given an appropriate affinity score, it is a simple matter to define the *best class label* for a collection of objects.

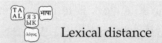

Lexical distance

Psychological motivation for the idea of a hierarchical distance between two concepts is given by the finding that it takes people longer to answer positively to the question "Is a robin an animal?" than to the question "Is a robin a bird?" (Collins and Quillian, 1969). However, other research shows that things are more complicated than this, with other factors playing a role; for example, how psychologically typical *robin* is as an example of the *bird* category (more typical than a *chicken*, in the opinion of many subjects; Miller 1998a, p. 32).

---

[7]That is, a function f that gets steadily smaller as $\text{dist}(w, h)$ gets bigger.

Definition 5 Let S be a set of nouns, let $\mathcal{H} = \bigcup_{w \in S} \text{Hyper}(w)$ be the set of hypernyms of S, and let $\alpha(w, h)$ be an affinity score function as defined in Equation (3.4). The *best class label* $h_{\max}(S)$ for S is the node $h_{\max} \in \mathcal{H}$ with the highest total affinity score summed over all the members of S, so $h_{\max}$ is the node which gives the maximum score

$$\max_{h \in \mathcal{H}} \sum_{w \in S} \alpha(w, h).$$

Since $\mathcal{H}$ is determined by S, $h_{\max}$ is solely determined by the set S, the affinity score α, and of course the lexical hierarchy being used (in this case, WordNet).

In the event that $h_{\max}$ is not unique (if there is a tie for first place), it is customary to take the most specific class label available.

The precise choice of class labelling algorithm depends on the functions f and g in the affinity score function α of equation (3.4). There is some tension here between being correct and being informative: correct but uninformative class labels (such as *entity*) can be obtained easily by preferring nodes high up in the hierarchy, but since our goal in this work is to classify unknown words in an informative *and* accurate fashion, the functions f and g have to be chosen to give an appropriate balance.[8] Similar trade-offs between being generally correct and specifically accurate arise in many other machine learning problems.

After a variety of heuristic tests, the function f was chosen to be

$$f = \frac{1}{\text{dist}(w, h)^2}, \qquad (3.5)$$

where for the distance function $\text{dist}(w, h)$ we used the simple method of counting the number of minimal (one-step) $\sqsubseteq$ relationships in a chain between w and h (defining the distance from any node to itself to be 1 rather than 0, to avoid dividing by zero). For the penalty function g we used the constant $g = 0.25$.

The net effect of choosing the reciprocal-distance-squared and a small constant penalty function was that hypernyms close to the concept in

---

[8] Again, we find Aristotle showing the way in the following guideline:

The man who gives an account of the nature of an individual tree will give a more instructive account by mentioning the species *tree* than by mentioning the genus *plant*. (*Categories*, Ch. 5).

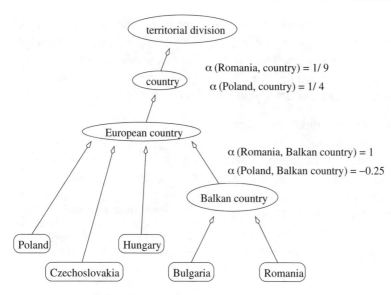

FIGURE 3.8 The class labelling algorithm at work.

question received magnified credit, but possible class labels were not penalized too harshly for missing out a node. This made the algorithm simple and robust to noise but with a strong preference for detailed, informative class labels.

Example 4 For the European countries in Figure 3.1, a summary of the behaviour of the class labelling algorithm is shown in Figure 3.8, where the set S of input nouns is {*Poland, Czechoslovakia, Hungary, Bulgaria, Romania*}.

The chain from *Bulgaria* to *country* is

Bulgaria ⊑ *Balkan country* ⊑ *European country* ⊑ *country*,

and since there are 3 links in this chain, we get the affinity score

$$\alpha(\textit{Bulgaria}, \textit{country}) = \frac{1}{3^2} = \frac{1}{9}.$$

The affinity score $\alpha(\textit{Bulgaria}, \textit{Balkan country})$ is just $\frac{1}{1} = 1$, which is much bigger than $\frac{1}{9}$ because *Balkan country* is a much more specific class label for *Romania* than just *country*. However, since *Balkan country* is not a hypernym of *Poland*, the affinity score $\alpha(\textit{Poland}, \textit{Balkan country})$ is the constant penalty function $-g = -0.25$. Also, because there is no *Balkan country* node intervening between *Poland* and *European country*, the affinity score $\alpha(\textit{Poland, country})$ is $\frac{1}{2^2} = \frac{1}{4}$. Adding together these affinity scores for all five of the countries in our input set S, we get the class labels and total affinity scores in Figure 3.1, the winner being the node *European country* with a total score of $3 \times \frac{1}{1} + 2 \times \frac{1}{4} = 3\frac{1}{2}$.

This example illustrates how dependent we are on the way knowledge is arranged in WordNet and the decisions that went into building it. It is largely a matter of historical convention that a few of the European countries are described collectively as "the Balkan countries" — we could just as easily have had a node for *Eastern European Country*, in which case this node would hopefully have been chosen as the best class label. The fact that some possible concepts are named and others are not contributes to a general issue with using taxonomies — the links in a taxonomy vary widely in the amount of information they give us. For example, the relationship *rabbit ears* ⊑ *tv-antenna* tells us much more than *sewing machine* ⊑ *machine*. In geometric terms, we might say that the *distance* between *rabbit ears* and *tv-antenna* is much shorter than the distance from *sewing machine* to *machine*. Resnik (1999) addresses this issue by computing the *information content* of each node, and hence the

 Inverse square laws

Equation 3.5 in an example of an *inverse square law* — a law which states that the attraction between two particles is proportional to the reciprocal of the square of the distance between them. The most famous such law was developed by Sir Isaac Newton (1642-1727) to describe the gravitational attraction between two particles or planets.

Newton realized that the exponent in the equation (in this case the number 2) is geometrically determined by the *dimension* of the space in which the mutual attraction takes place — in a 3 dimensional world, the gravitational force at a distance r from a planet is spread over the surface of a sphere, and since the surface area of this sphere is proportional to r^2, the force on each part of the sphere is proportional to $\frac{1}{r^2}$.

Efforts to determine the most suitable exponent in equation 3.5 could thus shed light on the number of dimensions which would be appropriate if a taxonomy was to be represented spatially. I would like to thank my father Kit Widdows for raising this point.

information content in each link, leading to a general method for computing distances and similarities in a taxonomy. Ways of measuring distances or similarities between concepts are very important, and will be the subject of Chapter 4.

Example 5 The least upper bound or *join*

One of the simplest ways to attach a class label to a list of objects in a taxonomy is to select the lowest node that subsumes *every* object in a set S. In more detail: we say that the node p is an *upper bound* for S if $s \sqsubseteq p$ is true for all $w \in S$, and p is a *least upper bound* if for any other upper bound q, we have that $p \sqsubseteq q$.

This node p is simply called the *join* of S (because it's the place where all the branches going from elements of S back to the root node join together) and is a crucial concept in lattice theory. In Chapter 7 we shall describe aspects of lattice theory, their role in logic, and the way these are used to describe linguistic concepts (Ganter and Wille, 1999).

In practice, the class labelling algorithm of Definition 5 can be tuned to give the join of its arguments by setting $\alpha(w, h) = 1$ for $w \sqsubseteq h$, and $g(w, h) = \infty$ (an infinite penalty) for $w \not\sqsubseteq h$. This ensures that if a node fails to subsume any one of the elements of S, it can never be chosen as a class label.

Since this particular class labelling function is so well-motivated by logical and theoretical considerations, why haven't we used it in practice? After all, it would give exactly the same answer for the European countries in Example 4. The answer is that it is such an exact method that there is virtually no margin for error. Even if all the objects of interest are represented in the taxonomy (or we choose to ignore those which aren't), this algorithm tends to give very general class labels such as *entity, something* for most sets, because by definition any class label has to subsume all of the input concepts. Another way of putting this is to say that in its search for completeness, this algorithm is far too sensitive to outliers in the data.

If we wanted to make sure that every class label subsumed *all* of the input words, the best class label we could have found for our list of countries would have been **geographical area** — technically accurate, but much less meaningful, less robust and in a sense less human than the class labelling algorithm we used in practice.

To give more meaningful answers we are forced to compromise and go with the flow a bit more. This is another of the many things at which humans naturally excel without even noticing it, and computers do not. In natural decision-making situations, if most of the evidence points to one conclusion we're often willing to follow it, even if there are reasonable doubts, provided we've weighed things up sensibly. Reasoning techniques such as the class labelling algorithm in Definition 6 attempt to model some of this robustness — though as we've seen, they still require careful tuning by humans.

Wider Reading

The main wider reading I recommend for this chapter is to delve into the writings of Aristotle, many of which are contained in editions such as that of McKeon (1941), and can alternatively be found on the internet in many places including the MIT classics archive.[9] A good way to begin is to read the *Categories*, which begins to describe a taxonomy of objects and their properties, and the ideas of species, genera and differentiae, in a very readable fashion. As an extra reward, in the very first line you'll see that the modern term **unequivocal**, the darling of so many politicians and pundits, is a cumbersome double negative with the same meaning as **univocal**.

Unlike Euclid's *Elements*, which was written as a college textbook, many of Aristotle's surviving writings are considered by modern scholars to be more like lecture notes or collected sayings than instruction manuals for students to read directly. A good primer or commentary can be an invaluable help in finding ones way through Aristotle's dense prose — I recommend the book by Barnes (2001), which is easy to read and small enough to fit in your pocket, both of these benefits enabling students to feel that they're getting to grips with Aristotle without a sense of inadequacy or pain.

Darwin's *Origin of Species* (1859) is another must-read for those who wish to understand the taxonomic approach. In particular, chapters 4 and 5 describe the principle of variation and examine the way in which variation and competition will lead to different species developing from an original genus. The mathematics behind this process is described

---

[9]classics.mit.edu/Aristotle/

in quite a sophisticated fashion, and many of the principles are also relevant to studying the way word senses vary and evolve (for example, on the whole more frequent words have more senses and more tendency to admit new variations, just as large genera are likely to produce more variation and hence more species).

The mathematics behind hierarchical structures and relationships in general will play a major role throughout the rest of this book. For an introduction to trees in terms of ordered sets and lattices see Davey and Priestley (1990, Ch 1): for an introduction in terms of graphs see Bollobás (1998, Ch 1).[10] Properties of relations (such as symmetry, transitivity, etc.) and their relation to (hierarchical) orderings are described in detail by Partee et al. (1993, Ch 3).

Directed graphs (sometimes called *digraphs*) are introduced clearly by Chartrand (1985, §1.6). Those interested in the idea of the World Wide Web as a directed graph may consult Kleinberg et al. (1999) and Kleinberg and Lawrence (2001), as well as reading about the *PageRank* algorithm (Brin and Page, 1998).

Many of the chapters (particularly 1, 2, 3 and 5) in the yellow WordNet book (Fellbaum, 1998b) are strongly recommended, since so much information they contain is freely used throughout this book. The introduction to hyponymy given in this chapter might suggest that hyponymy is a fairly simple relationship to define and to understand: this is far from the case. Several varieties of hyponymy and their interrelations are discussed by Cruse (2002). A readable summary of lexical relationships such as synonymy, hyponymy and meronymy is given by Aitchison (2002, Ch 8,9).

There is a baffling wealth of literature concerning semantic networks: good places to start include Lehmann (1992) and Sowa (2000). In discussing the *Cyc* database, Lenat and Guha (1990) introduce the minefield of potential pitfalls and mistakes (such as confusing an individual member relationship $a \in b$ with a hyponymic relationship $a \sqsubseteq b$). More linguistic issues can be found in Evens (1988), a collection which describes different ways semantic networks have been used to represent specifically *lexical* information, and how semantic networks can be combined with dictionaries and thesauri. (A notable variation between approaches is in the number of different *types* of semantic

---

[10]Note that in terms of graph theory, a tree need not be a *directed* graph.

relationship which different authors find to be necessary in their systems, answers ranging from 3 to precisely 53 to over 100.) The links in a knowledge base have been encoded by hand (as with the *Cyc* knowledge base (Lenat and Guha, 1990)),[11], mined from a dictionary or another machine readable resource (as extensively investigated in the MindNet project[12] (Dolan et al., 1993)), and learned directly from free text (Hearst, 1992). Readers wishing to know how taxonomies have been developed in languages other than English might well begin by reading about EuroWordNet (Vossen, 1998).

The early work of Quillian (1968) sets the tone for much of what has followed in subsequent decades: in addition, this work provides some of the first automatic techniques for measuring semantic similarity between words, and this forms a valuable grounding for understanding Chapter 4 of this book. Quillian's network is *not* a hierarchy in the sense of having privileged root nodes, but takes the alternative approach of *spreading activation*, where each concept can be activated as its own root concept, whereafter its relations are activated recursively to form a tree around it.

The class labelling algorithm in this chapter first appeared in Widdows (2003c), where it was used for the express purpose of enriching a taxonomy such as WordNet. Widely different results were obtained for common nouns, proper nouns, and verbs, reflecting to some extent the degree to which a hierarchical model is appropriate for describing the meanings of these different categories. Earlier work in this area includes that of Hearst (1998) and Caraballo (1999).

The task of finding a part of a taxonomy in which a set of words are most concentrated (which we have called class labelling) has been comparatively widely studied. Two good versions are

- Agirre and Rigau (1996) use a measure called *conceptual density* for finding the correct meaning of an ambiguous word in a particular context.
- Li and Abe (1998a) use the *minimum description length*, an idea from information theory, to model the selectional preferences of verbs (for example, to discover that if something *flies* it is usually an *aircraft*, an *insect* or a *bird*). This is a more sophisticated problem (and solution)

---

[11]www.cyc.com with an extensive Open Source version www.opencyc.org
[12]research.microsoft.com/nlp/Projects/MindNet.aspx

than our simple class labelling technique, because the best class of objects may be from a combination of different parts of the taxonomy rather than just one branch.

Information theory has already been mentioned a couple of times, in describing the work of Li and Abe (1998a) and Resnik (1999), which uses information theory to measure the similarity between two nodes in a taxonomy. Information theory is an increasingly important part of the computational linguist's toolbox, and though a description of this subject is beyond the scope of a book on geometry, readers will certainly get more out of this book if they are familiar with an introduction such as that given in Manning and Schütze (1999, §2.2), which is tailored for linguistic applications. Shannon's (1948) founding paper on information theory remains a very rewarding read, contains several examples from language, and is widely available in several reprinted forms. In a nutshell, if (for example) someone tells you that a word in English begins with an *x*, that gives you a lot more *information* than telling you that it begins with an *s*, because on the whole *x* is a much less probable first letter for an English word than *s*. Information theory gives a precise way of formalizing and measuring this information content.

4

Measuring Similarity and Distance

So far we have considered sets of objects, and relationships between these objects such as the symmetric relationships $a \leftrightarrow b$ in the graph model of Chapter 2, and the asymmetric or vertical relationships $a \sqsubseteq b$ of the concept hierarchies in Chapter 3. We have started to feel our way around a graph or a hierarchy, following chains of relationships and learning about the shape of the space in the vicinity of a few interesting example words.

We have still said very little about *measuring* these relationships. In building the graph (Section 2.4), we considered how much weight should be given to each link, and gave some rules of thumb for ranking links. When using WordNet, we wanted to give greater significance to the relationship *oak* $\sqsubseteq$ *tree* than to the relationship *oak* $\sqsubseteq$ *living thing*, because *oak* is *closer* to *tree* that it is to *living thing*.

In order to obtain more general results, we need to be able to compare relationships more systematically. One of the most tried and tested mathematical techniques for doing this is to measure the *distance* between two points. Conversely, we might try to measure the *similarity* between two points (small distances corresponding to large similarities and large distances corresponding to small similarities).

This chapter introduces some of the ways that have been used to measure distances and similarities in mathematical spaces such as graphs and hierarchies. The most standard distance measures in mathematics are called *metrics*, which must satisfy certain conditions or *axioms* (such as being symmetric).

However, we must also pay great care to testing whether these mathematical techniques are actually appropriate when we're dealing with language. For example, we shall see that some of the properties of metrics are not always ideal for describing distances between words and concepts, and this chapter presents a few case studies in what can go wrong if we are not very careful and sensitive to the goal of our work, which is not using mathematical ideas of distance but inferring similarity of meaning in natural language.

Measuring similarity between words can enable us to build whole classes of words with similar semantic properties — sets of words which are all *tools* or all *musical instruments*, for example — and we demonstrate that the graph model of Chapter 2 can be used for this purpose very successfully, provided we show great care in the presence of ambiguous words. Ambiguous words can sometimes behave like *semantic wormholes*, accidentally transporting us from one area of meaning to another. But we can also use this effect positively, because finding those words which have this strange wormhole effect when measuring distances helps us to recognize which words are ambiguous in the first place.

4.1 Distance functions

This section introduces some standard ideas of distance in mathematics, and in particular metric functions and metric spaces. Suppose we have a set A. Then a *distance function* on A is a mapping

$$d : A \times A \to \mathbb{R},$$

which assigns to the pair $(a, b) \in A \times A$ a real number $d(a, b) \in \mathbb{R}$ called the *distance* between a and b. Like many of the geometric ideas we've encountered, distance is a property not of a single point, but a relationship between a *pair* of points.[1]

Needless to say, not just any function d will make a good distance function — there are a few more properties that $d(a, b)$ should arguably possess. For example, if c lies on the way from a to b, then the distance

---

[1] Even in day-to-day speech when we omit to mention both points, this is still true — if only one point is mentioned then the other, by default, is the one where the speaker or the listener is currently situated. Breaches of this convention can be deliberately humourous, for example

"... In South America."

"That's abroad, isn't it?"

"Well it all depends on where you're standing."

(From Spike Milligan's *Goon Show* radio script, *The affair of the Lone Banana*.)

from a to c should be *less* than the distance from a to b. This is the reasoning that lies behind the intuition that, in the WordNet taxonomy, $d(oak, tree)$ should be less than $d(oak, living\ thing)$, because the chain from *oak* up to *living thing* passes through the *tree* node.

A simpler prerequisite for a distance function is that if two points are identical then there is *no* distance between them — the distance from any place to itself is zero, or $d(a, a) = 0$ for all $a \in A$. On the other hand, if two points *aren't* the same then the distance between them must be greater than zero — we have no teleportals in our space as yet. Another rule of thumb is that the distance from a to b should be the same as the distance from b to a, so that $d(a, b) = d(b, a)$ for all $a, b \in A$. This gives us the following rules or *axioms* for a distance function to be a *metric*:

Definition 6 Let A be a set and let $d : A \times A \to \mathbb{R}$. The function d is a *metric* on A if and only if the following axioms hold.

1. $d(a, b) \geq 0$ for all $a, b \in A$, and $d(a, b) = 0$ if and only if $a = b$.
2. $d(a, b) = d(b, a)$ for all $a, b \in A$. (So d is symmetric.)
3. For all $a, b, c \in M$, $d(a, c) \leq d(a, b) + d(b, c)$.

In this case d is called a *metric* on A and (A, d) is called a *metric space*.

This may seem like a lot of rules, but in isolation each is quite simple. Axiom 1 states that the distance between two different places must be positive and Axiom 2 rules out the existence of 'one-way streets.' Axiom 2 is clearly another *symmetric* rule, saying that $d(a, b)$ stays the same if we interchange a and b.

Axiom 3 is particularly interesting — it says that you can't make a and c any closer together by taking a short-cut through b, wherever a, b and c are. This rule is sometimes known as the *triangle inequality* — for every triangle abc, the length of the longest side $d(a, c)$ is *never* greater than the sum of the lengths of the two shorter sides $d(a, b) + d(b, c)$ (Figure 4.1).

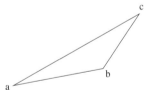

FIGURE 4.1 The length ac cannot exceed the sum of the lengths ab and bc.

The three metric space axioms, and the term 'metric space' (*Metrischer Raum*) itself, were introduced by Felix Hausdorff (1914, Ch 6), one of the founding fathers of modern set theory and topology, and this gave a very successful formal method for defining spatial ideas such as connectedness and convexity. Tragically, Hausdorff committed suicide in 1942, together with his wife and her sister.

Example 6 Shortest paths in a graph

The simplest way to define a metric in a graph G is to say that, for two nodes $a, b \in G$, the distance in between them $d(a, b)$ is the number of links in the shortest path from a to b. It is not hard to see that this satisfies the three Axioms in Definition 6, and so d is a metric on G.[2]

The shortest path is a good metric for some applications, such as working out the cost of shipping a cargo from one point to another in a distribution network. However, it doesn't always work so well for the graph of words in Chapter 2 — in Figure 2.6, it would predict that the distance from *cheese* to *margarine* (2 links) is the same as the distance from *cheese* to *limestone* (also 2 links). It would be beneficial to find some way of saying 'it looks as though the link from *cheese* to *chalk* is pretty dubious — don't put too much faith in it for tracing shortest paths between concepts.' A pretty successful method for doing precisely this will be given later in this chapter, motivated by the theoretical considerations we will discuss along the way.

4.1.1 Pythagoras' theorem and Euclidean distance

The single most widely-used metric function remains the *Euclidean distance* function which is derived from Pythagoras' theorem,

> In right-angled triangles the square on the side opposite the right angle equals the sum of the squares on the sides containing the right angle.[3]

In Figure 4.2, this gives the equation $(d(a,b))^2 + (d(b,c))^2 = (d(a,c))^2$.

---

[2]One whimsical application of this metric is in the computation of *Erdös Numbers*. The mathematician Paul Erdös (1913-1996) was a prolific author and collaborated very widely with hundreds of coauthors. These coauthors are said to have an Erdös number of 1. *Their* coauthors are said to have an Erdös number of 2, etc. Thinking of the collection of all authors as nodes in a graph, and placing a link between two authors if they were coauthors on any paper, the Erdös number of any author a is the distance $d(\textit{erdös}, a)$ in this graph using the shortest path metric. Similar fun is had in the film industry with the *Kevin Bacon game* (http://www.cs.virginia.edu/oracle/), and in social networks generally with ideas like '6 degrees of separation.'

[3]This theorem appears in Euclid's *Elements of Geometry* (Book I, Proposition 47), and (as is common with naming conventions in mathematics) is attributed to Pythagoras because he is reputed to be the first *Westerner* to make the discovery. The theorem was clearly known throughout many cultures much earlier, the Babylonians in particular

Far from being old-hat, Pythagoras' theorem is invaluable in modern applications because it allows us to work out arbitrary distances in terms of easily measured ones. Suppose we know the x and y coordinates of the points $a = (x_1, y_1)$, $b = (x_2, y_1)$ and $c = (x_2, y_2)$ (Figure 4.2). Then a and b only differ in their x coordinates, so we can easily measure $d(a, b) = x_2 - x_1$, and similarly, $d(b, c)$ is easily measured to be $y_2 - y_1$. We now use Pythagoras' theorem to deduce that

$$(d(a,c))^2 = (d(a,b))^2 + (d(b,c))^2 = (x_2 - x_1)^2 + (y_2 - y_1)^2,$$

or just

$$d(a,c) = \sqrt{(x_2 - x_1)^2 + (y_2 - y_1)^2}. \qquad (4.6)$$

Is is not difficult to verify that this distance function satisfies all three axioms in Definition 6, and so Pythagoras' theorem actually gives a *metric* on the plane, which is most often called the *Euclidean distance*, because it is the distance measure which is used implicitly in Euclidean geometry to compare (for example) the lengths of lines.

The trick is to represent a point by a number of coordinates, and then Pythagoras' theorem tells us how to manipulate these coordinates to give the distances between points. We've gone into some detail here because it will be very useful for us to generalize this metric to systems with more than two coordinates in Chapter 5.

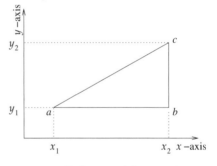

FIGURE 4.2 Pythagoras' theorem in x-y coordinates

There are other metrics that can be used to measure distances just in the plane, before we even consider other sets of points. For example, the city-block or *Manhattan* metric measures distances as you actually have to travel them in a city laid out in a grid — if you can't go diagonally, then to go '3 blocks west and 4 blocks north' that's precisely the route you have to take, travelling a distance of 7 block-widths. In Figure 4.2, this would give the distance

---

demonstrating an in-depth knowledge of the number theory used for generating Pythagorean triangles (Boyer and Merzbach, 1991, p. 34). Van der Waerden (1985) suggests that the legend of Pythagoras celebrating his discovery with a sacrifice of (one or several) oxen is an anachronism, since the Pythagoreans themselves were strict vegetarians, and that this legend could derive from an earlier Indo-European priestly tradition. However, the rigour of the *proof* of the famous theorem in Greek geometry was probably a milestone.

$d(a,c) = (x_2 - x_1) + (y_2 - y_1)$. Such metrics are described in detail by Hausdorff (1914, p. 116), and more accessibly by Gärdenfors (2000, p. 18).

Many geometric concepts actually depend on choosing a metric. For example a *straight line* is normally defined as 'the shortest distance between two points' for which we need an idea of 'shortest distance.' That different ways of measuring distance give different shortest distances makes the notion of 'straight line' adaptable to its context rather than absolute and fixed.

4.2 Similarity Measures

A *similarity function* or *similarity measure* is the converse of a distance measure. For a set A, a good similarity function sim : $A \times A \to \mathbb{R}$ should give a large value for $\text{sim}(a,b)$ if a and b are similar objects, and a small value if a and b are very different from one another. Some standard measures of similarity, and the various motivations underlying them, are given by Manning and Schütze (1999, §8.5).

In spatial language, similarity corresponds with proximity and dissimilarity corresponds with distance. If two points are very distant from one another, they will be regarded as dissimilar, and if they are very close together, they will be regarded as similar. Thus a similarity measure is in a sense the opposite of a distance measure, and there are simple ways for transforming between them.

If $d : A \times A \to \mathbb{R}$ is a distance measure, then a simple way to obtain a similarity measure sim : $A \times A \to \mathbb{R}$ would be to set

$$\text{sim}(a,b) = \frac{1}{d(a,b)}, \tag{4.7}$$

which is used for example to transform a *resistance* into a *conductance* in physics. Assuming that d has the three properties of a metric, we must have $d(a,a) = 0$, in which case $\text{sim}(a,a)$ would become undefined (or would tend to infinity). If we want objects to be infinitely self-similar, this would be fine, but instead we might prefer to set

$$\text{sim}(a,b) = \frac{1}{1 + d(a,b)}, \tag{4.8}$$

which would ensure that $\text{sim}(a,a) = 1$ and that for any other object b, $\text{sim}(a,b) < 1$. If desirable, this may allow us to interpret $\text{sim}(a,b)$ as the probability that a and b are interchangeable. In the end, there is probably no canonical 'best' transformation between similarity functions and distance functions, but a variety that will be good in different situations.

The psychological and practical value of similarity as a concept has been strenuously attacked (and defended) for a number of reasons (Gärdenfors, 2000, p. 109). One of the main problems is that globally defined similarities are not sensitive to context, so the judgment that two things are similar or dissimilar can easily be taken out of context. To some extent, this can be catered for by tuning a similarity (or distance) function to give more importance to different variables in different situations (Gärdenfors, 2000, p. 132). In

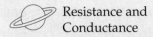

Resistance and Conductance

The relationship between the *resistance* and the *conductance* of a link in an electrical circuit (Bollobás, 1998, p. 42) is a typical example of the transformation between a measure of 'distance' and a measure of 'similarity' in the physical sciences. Resistance R measures the difficulty with which an electric current will pass, conductance C measures the corresponding ease, and the two are related by the simple reciprocal relationship

$$C = \frac{1}{R}$$

Figure 4.2, this could be implemented by rescaling the x-axis or the y-axis. However, until we have a sure-fire way of *automatically* adapting a similarity or distance to cope with variations in context, this remains a practical problem. For example, a similarity thesaurus (page 44) may know that *tree* and *bush* are similar: but suppose you want to find information about *trees* in the sense of the hierarchies of Chapter 3. Then if the thesaurus tells the search engine that documents with the keyword *bush* are probably relevant as well, it would actually damage the search engine's performance.

In spite of this objection, notions of similarity and distance contribute a tremendous amount to practical language engineering, especially in situations where we are trying to find a best choice from a number of candidates. Choosing the most relevant documents in a search engine, the best translation from a number of suggested translations for a particular sentence, or the most likely sense of an ambiguous word in a particular context: all these are processes where at some point a measure of distance or similarity is used to judge which of the possible choices is closest to being an ideal solution.

4.2.1 Cosine similarity between points on the unit circle

This section describes one of the most widely used measures of similarity. It is very important because it generalizes very easily to any number of dimensions, without significant alteration. This will enable us to calculate similarities between points with *any* number of coordinates (that is, in spaces of any dimension) throughout the rest of this book. The good news is that it's very simple to understand, takes only moments to program on a computer (however many points and dimensions you have), and runs very efficiently. The bad news (of course) is that no such simple measure could be expected to perform well in all possible situations. Nonetheless, cosine similarity has been a remarkably versatile and popular measure in many situations.

For now, we will only consider points on the unit circle, i.e. the circle whose distance from the origin is one unit. Dealing only with points on this circle is a way of focussing our interest on the *directions* of a and b (the directions in which their arrows are pointing), rather than their *magnitudes* (their distances from the origin point $(0,0)$). Let a and b be two points lying on the unit circle, whose a and b coordinates (x_1, y_1) and (x_2, y_2) are shown in Figure 4.3. In

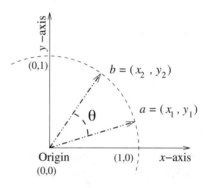

FIGURE 4.3 If two points are in similar parts of the circle, the angle in between them is small

this case the angle θ in between the points is a good measure of the *distance* between them. We now invert this into a measure of *similarity*.

If the two points are close together, then the angle in between them is small, and we might say that they are fairly similar to one another: if they are *exactly* the same point, then we shall say that their similarity is equal to 1. On the other hand, suppose that the points a and b are at right angles (± 90 degrees) to one another, so that their arrows are perpendicular or *orthogonal* to one another. Then we might be tempted to say that they have nothing in common at all, so that their similarity would be equal to 0. At the other extreme, if a and b were completely opposite (± 180 degrees) to one another, we will say that their similarity

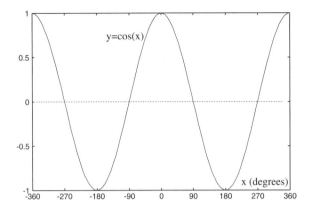

FIGURE 4.4 The cosine function or cos wave.

is equal to -1, and if we go any further round the circle than 180 degrees we start going back towards our initial point.

Two pieces of good fortune enter the scene for us. Firstly, the cosine function from trigonometry provides an ideal transformation from the angle θ to the desired similarity measure (Figure 4.4). The second is that we don't even have to work out the angle itself to calculate its cosine. For two points a and b on the unit circle (Figure 4.3), all we have to do is multiply each of the coordinates for the point a with the corresponding coordinate for the point b, add the results together, and we get *the same answer* (see Jänich, 1994, p. 31).

Definition 7 Cosine similarity[4]

Let two points a and b lying in the unit circle have coordinates $a = (x_1, y_1)$ and $b = (x_2, y_2)$. The *cosine similarity* of the points a and b will be written $\cos(a, b)$ and is given by the formula

$$\cos(a, b) = x_1 x_2 + y_1 y_2.$$

---

[4]There are a variety of names and notations for cosine similarity, which is closely related to the scalar product of two vectors (see Chapter 5). The notation $\cos(a, b)$ is that of Hamilton (1847) in his work on quaternions, in which vectors were pioneered, and is also widely used in computational linguistics, partly because of its unambiguous meaning when dealing with vectors containing information about words. The notation $\sim(a, b)$ is also widely used (Jurafsky and Martin, 2000, p. 650), though this is also used for other similarity measures.

Thus cosine similarity gives us a very efficient, convenient and (as we shall see) general method for taking the coordinates of two points on a circle, multiplying these, and adding the answers together to give a single number between 1 and −1 which tells us whether these points are similar, opposite or perpendicular.

4.2.2 Similarity and distance in hierarchies

A variety of robust and flexible measures of similarity and distance have been developed which take as input a pair of points a and b in a hierarchy or taxonomy, and return a similarity $\text{sim}(a,b)$ or distance $d(a,b)$ derived from the structure of the taxonomy itself.

Some methods are overtly geometric, such as that of Leacock and Chodorow (1998), which gives a similarity $\text{sim}(a,b)$ determined by the length $d(a,b)$ of the shortest path from a to b.

> **WordNet Similarity measures**
>
> An implementation of the most prominent similarity measures for taxonomies, applied to WordNet, is available as a collection of Perl modules, written and maintained by researchers at the University of Minnesota (Patwardhan et al., 2002). The module is freely available from sourceforge.net/projects/wn-similarity/ and can be installed alongside a compatible version of WordNet.

Length is defined by using the graph structure of the hierarchy and counting the number of links in a chain from a to b, as in Example 6 (page 101). This is then converted to a similarity measure using the negative of the logarithm function as a transformation, so that

$$\text{sim}(a,b) = -\log\left(\frac{d(a,b)}{2D}\right),$$

where the constant factor D is the maximum depth of the taxonomy. Another geometric measure of similarity is given by Hirst and St-Onge (1998), which takes into account not only the length of the shortest path, but the number of times this path changes direction.

A more probabilistic method was designed by Resnik (1999), which defines similarity in terms of information content, and is notably the first similarity measure to combine predefined information from a hand-built taxonomy with empirically gathered statistics from a large corpus. To find the similarity between two nodes a and b, first find the lowest node

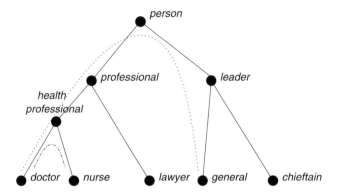

FIGURE 4.5 A fragment of the WordNet hierarchy, from which we infer that *doctor* is more similar to *nurse* than to *general* because the path between them is much shorter, or alternatively because the join node *health professional* is much more specific than the join node *person*.

which subsumes both a and b, which is called the *join* of a and b, written $a \vee b$ (the join idea was used on page 93 and is described properly in Chapter 8). For example, in Figure 4.5, the join of *general* and *chieftain* is *leader* and the join of *lawyer* and *doctor* is *professional*. Secondly, find the frequency with which the words in this branch of the taxonomy (all the words subsumed by the node $a \vee b$) are used in an actual corpus. This gives a measure of the probability that any word chosen at random from the corpus will refer to one of the concepts in the branch defined by $a \vee b$ — call this probability $Pr(a \vee b)$. Finally, we measure the *information content* of this node, which is the number of bits which on average it takes to encode this information, namely $-\log_2(Pr(a \vee b))$. This gives the final similarity score

$$\text{sim}(a, b) = -\log_2(Pr(a \vee b)).$$

The intuition behind this measure is as follows. Consider the pairs (*doctor, nurse*) and (*doctor, general*), shown in Figure 4.5. In WordNet, the join of *doctor* and *nurse* (written *doctor* ∨ *nurse*) is the node *health professional*. On the other hand, the join *doctor* ∨ *general* is the much less specific node *person*. Being told that an object a is a *person* is much less informative than being told that a is a *health professional*, and for this reason we infer that sim(*doctor, nurse*) is greater than sim(*doctor, general*). Other information-theoretic measures of similarity in a taxonomy include those of Jiang and Conrath (1997) (a distance measure rather than a similarity measure) and Lin (1998b).

In practice, all of these measures are complicated by the appearance of ambiguity — the word *doctor* appears in several places in WordNet, attached to different nodes with hypernyms such as *theologian* and *scholar* as well as *health professional*. Because of this, the similarity or distance between *doctor* and any other node in the hierarchy becomes multiply defined. However, this can also be used as an opportunity to *resolve* ambiguity — if you encounter the phrase "doctor and nurse," you can reason that the *doctor* in question probably refers to a physician rather than to a scholar or a theologian, because this meaning of *doctor* is the most similar to the meaning of *nurse*, a method implemented by Resnik (1999). This is a good example of the way ambiguity can make measuring similarities or distances more complicated, but at the same time these very same complications can be used to give extra insight about the behaviour of ambiguous words. We will return to this theme at the end of this chapter.

With all these similarity measures in the literature, it begs the question "which is the best one to use?" Needless to say, there is no single answer: not surprisingly, since there is really more than one question here. One way to phrase the question more precisely is to ask which of the measures performs best on a particular predefined task, such as automatically correcting spelling mistakes (Budanitsky and Hirst, 2001) or resolving word-sense ambiguity (Patwardhan et al., 2002). Both these studies found that the measure of Jiang and Conrath (1997) was particularly effective, though the reasons *why* this measure performed best in these experiments are still unclear.

Even then, there is as yet no way to guarantee that a similarity measure which performed well on one task would perform well on another. For a search engine or a question-answering system, a more specific piece of information may be fine: if there is a question "How do quadrupeds normally move?" then a sentence telling us how a *horse* moves will be at least informative, even if not the whole story, so a measure which told the system that a *horse* and a *quadruped* are similar concepts might be useful. On the other hand, giving a very specific classification of which one is unsure when automatically *building* a taxonomy could be actively misleading — if a *cow* was classified as a kind of *horse* instead of a kind of *quadruped*, one might go on to infer that a *cow* can be *ridden*, since horses have this property.

4.3 Which distance axioms are appropriate for word meanings?

Another way to compare different measures of similarity or distance is to examine the principles upon which they are built, and see if they correspond well with our intuitive grasp of reality. This may have its own drawbacks — one could be convinced that a method *should* work perfectly, but especially when dealing with human language, there are always surprises in store. However, a careful analysis of the mathematical and psychological assumptions underlying a particular measure can be an invaluable guide to building a successful algorithm or a workable theory. As we shall see in the rest of this chapter, it is possible at least to work out which assumptions are *not* appropriate, and to obviate the damage that such assumptions might cause. This process provides a typical example of the sort of issues that arise when considering which mathematical models should be used to describe a real situation, and as such will hopefully provide the reader with an interesting case study of this part of the scientific process.

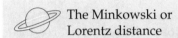 The Minkowski or Lorentz distance

One of the most notable distance measures which does not obey the metric axioms is the *Minkowski* or *Lorentz* distance, used in Einstein's Special Theory of Relativity to give the distance between two points in spacetime. Let ∇x (the change in x), ∇y, ∇z and ∇t be the change in a particle's spatial (x, y and z) and time (t) coordinates as it moves from point a to point b. Then the distance between a and b is given by the equation

$$(d(a,b))^2 = \nabla x^2 + \nabla y^2 + \nabla z^2 - (c\nabla t)^2$$

the constant c being the velocity of light. Note the way the time coordinate t becomes intertwined with the spatial coordinates, but with a *negative* value, so that there are many points b with $d(a,b) = 0$. In this case the path from a to b is called *lightlike*, because only a beam of light can move fast enough to reach b from a. This contradicts the first metric space axiom which states that $d(a,b) = 0$ only if a and b are identical — it's not only in language or psychology that it pays to break mathematical rules!

Many features of full metrics are known to be inappropriate for describing the relationships between concepts. In psychology, this point was demonstrated in a classic work by Tversky (1977), and to some extent this section merely summarizes Tversky's objections.[5] Take a simple example — in mathematics, a metric function must satisfy the symmetric property $d(a,b) = d(b,a)$ (Definition 6, Axiom 2). But as Tversky points out,

> We say "the portrait resembles the person," rather than "the person resembles the portrait." We say "the son resembles the father" rather than "the father resembles the son." We say "an ellipse is like a circle" not "a circle is like an ellipse," and we say "North Korea is like Red China" rather than "Red China is like North Korea."

These examples are summed up by the following *prototype–variant rule*:

Observation 1 The variant is more similar to the prototype than the prototype is to the variant.

Since the ⊑ relationships in a taxonomy are *antisymmetric* rather than symmetric, we may well expect a symmetric similarity or distance function to be less than ideal. Taking the prototype–variant rule (Observation 1) seriously, we may instead suggest that the distance from a child node to its parent node should be *less* than the converse distance — so that, for example, $d(\textit{oak}, \textit{tree}) < d(\textit{tree}, \textit{oak})$. With this in mind, Faatz et al. (2001) design a distance function for taxonomies which deliberately assigns a shorter distance to steps *up* the tree than to steps *down* the tree, because "Finding superconcepts is more fault-tolerant than referring to senseless specializations." Again, it is also important to take the *purpose* behind a similarity or distance measure into account — the distance function of Faatz et al. (2001) is designed for enriching an ontology, and as we noted earlier, this is a task for which 'senseless specialization' may be much more damaging than for (say) information retrieval.

4.3.1 Prototypes and variants in the graph model

A little has already been said about weighting the different links in a graph of nouns (Section 2.4). It turns out that we can adopt a weighting scheme to give a similarity measure between linked nodes

---

[5]Even the first assumption that $d(a,b) \geq d(a,a)$ ('the thing most similar to an object is itself,' Definition 6, Axiom 1) is questioned by Tversky, since some identical stimuli are apparently easily confused.

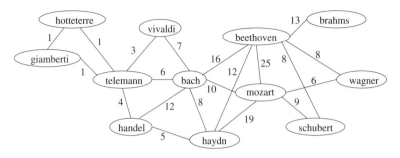

FIGURE 4.6 The observed frequencies of links between composers.

in the graph in ways which naturally incorporate the prototype-variant observation (1). Back in the graph model built from the BNC, consider the collection of composers in Figure 4.6. As well as a (slightly simplified) collection of nodes and links, this picture includes example link strengths, these being the numbers next to each link reflecting the number of times the link was observed in a coordination pattern in the BNC. If we take these raw numbers as a similarity score then we get values such as $\text{sim}(\textit{beethoven}, \textit{bach}) = 16$, which is of course symmetric because $\text{sim}(\textit{bach}, \textit{beethoven}) = 16$ also.

There are drawbacks to this scoring technique: in particular, frequently occurring words with many links will be regarded as strongly similar to many many words. For example, this would give $\text{sim}(\textit{mozart}, \textit{wagner}) = 6$, even though the two composers are from different historical periods and their music is of very different styles. At the other extreme, rare words will be awarded little similarity with any other words — for example, in Figure 4.6 we would only have $\text{sim}(\textit{hottenterre}, \textit{giamberti}) = 1$, even though they are both from the Baroque period and wrote music for the recorder. At the least, it seems unfair to say that *mozart* and *wagner* are 6 times more *similar* largely because they are both more *famous*.

There is a simple way round this problem — we can divide through by this fame factor to get a more even playing field. To do this, define the *weight* of each node in the graph to be the number of links it has (so a link of strength 16 can be thought of as 16 separate links and contributes this total to the weight score). We might write this score as $\text{wgt}(a)$, so that (for example) we have $\text{wgt}(\textit{mozart}) = 25 + 10 + 19 + 9 + 6 = 69$ and $\text{wgt}(\textit{hottenterre}) = 1 + 1 = 2$. If we divide the link strength score (which we will call $\mathrm{ls}(a,b)$) by this number we get the similarity score

$$\mathrm{sim}(a,b) = \frac{\mathrm{ls}(a,b)}{\mathrm{wgt}(a)} \qquad (4.9)$$

which gives the much more reasonable similarity scores

$$\mathrm{sim}(\textit{mozart, wagner}) = \frac{6}{69} \approx 0.087$$

and

$$\mathrm{sim}(\textit{hottenterre, giamberti}) = \frac{1}{2} = 0.5.$$

The step of dividing by some measure of total frequency to make things more fair is generally referred to as *normalization*. Similarity measures such as the one in Equation (4.9) have been used for a long time (going back to Bush and Mosteller, 1951) and can be motivated by considerations from basic set theory.[6] In probabilistic terms, we can also interpret $\mathrm{sim}(a,b)$ as the probability that a particle starting a random walk at the node a will next visit the node b.

We now note that our measure of similarity is *no longer symmetric*, because the weights of the nodes a and b in Equation (4.9) are not usually equal. For example, we have

$$\mathrm{sim}(\textit{telemann, bach}) = \frac{6}{15} = 0.4$$

whereas

$$\mathrm{sim}(\textit{bach, telemann}) = \frac{6}{59} \approx 0.102,$$

so that *telemann* is measured as being much more similar to *bach* than *bach* is to *telemann*. In this way, the normalization step brings the graph model into direct accord with the prototype–variant observation (1) — it's pretty fair to say that *bach*[7] is the very prototype of a *baroque composer*, and that *telemann* is a variant within this group, and we have the result that the variant is deemed more similar to the prototype than the prototype is to the variant. Note that this chain of reasoning can

---

[6]To make this transition from graph theory to set theory, think of the weight $\mathrm{wgt}(a)$ as corresponding to the size of a set X, and the link strength $\mathrm{ls}(a,b)$ corresponding to the size of a set $X \cap Y$, giving the similarity score $\mathrm{sim}(X,Y) = |X \cap Y|/|X|$.

[7]Referring to Johann Sebastian Bach (1685–1750): he had many sons, several of whom became distinguished composers in their own right, though are normally regarded as belonging to the early classical rather than the baroque period. The graph model does not as yet have any means of distinguishing the different individuals referred to as *bach*; this problem in disambiguation has been addressed by Mann and Yarowsky (2003).

be followed throughout this small graph, with **beethoven** and **mozart** being perhaps the most prototypical composers of them all.[8]

4.4 Transitive relationships

This section explores another of the key kinds of mathematical relationships, which turns out to be particularly relevant when inferring new similarities from known ones. Transitivity is also closely related to the triangle inequality (Axiom 3, page 100), so it makes particular sense to discuss it while we are exploring the extent to which the metric space axioms are appropriate for measuring distances between concepts. Just as there are well-founded objections to thinking of similarity between concepts

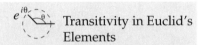

Transitivity in Euclid's Elements

The notion of transitivity for the relationship of *equality* in mathematics is very ancient, and Euclid deems it sufficiently important to list as his first common notion

Things which equal the same thing also equal one another.

In modern notation, this would be the argument that if $a = b$ and $b = c$, then $a = c$ also. This is true for things which are equal, but it does not always generalize to things which are *similar*.

as a symmetric relationship, there are deep (and possibly worse) problems that arise when assuming that a similarity relationship between concepts should be transitive. However, transitivity can also be used to extrapolate from partial experience, which is a vital part of computational linguistics: we will never encounter every possible piece of useful information in text, and so we have sometimes to infer information indirectly. Transitivity is a two-edged sword: we need to use it, and we need to use it with great care.

We have already encountered the concept of transitivity (page 77) when considering hierarchical ⊑ relationships. The basic assumption that hierarchical relationship are transitive allows us to reason (for example) that every *horse* is a *mammal*, every *mammal* is a *vertebrate*, therefore every *horse* is also a *vertebrate*. For a general relationship ∼, transitivity is defined as follows.

---

[8]As stated earlier (footnote to page 57), the simple heuristic rule of taking the top n neighbours of a given node (Widdows and Dorow, 2002) also modifies the graph so that it is no longer symmetric, and this modification is again in line with the prototype–variant observation.

Definition 8 Let $\sim$ be any relationship between mathematical objects. Then $\sim$ is said to be *transitive* if whenever $a \sim b$ and $b \sim c$, it also follows that $a \sim c$.

The word 'transitive' to describe such relationships was possibly first used by the 19th century mathematician and logician Augustus de Morgan, the idea being that in the statement $a \sim b \sim c$, the relationship $\sim$ 'transits' over b to give $a \sim c$. The triangle inequality (Axiom 3, page 100) is effectively a looser form of transitivity — if a is close to b and b is close to c, it places a maximum limit on how far a can be from c, roughly saying "if a is near b and b is near c, then a is not very far from c." Thus it makes sense to analyze both transitivity and the triangle inequality as parts of the same package.

Whether a relationship is transitive is *not* determined by whether it is symmetric or antisymmetric or neither — it is a new and independent concept. Some symmetric relationships are transitive and others are not; some antisymmetric relationships are transitive and others are not.

Transitivity and Meronymy

The meronymy (PART_OF) relationship is not always transitive — for example, a *door handle* is part of a *door* and the *door* may be part of a *room*, but we would not normally say that the *door handle* is part of the *room*.

In some cases this is because there are different kinds of meronymy — the physical PART_OF relationship (e.g. *tail* PART_OF *wolf*) and the MEMBER_OF relationship (e.g. *wolf* MEMBER_OF *pack*, or *wolf* $\in$ *pack*) may both be kinds of meronymy, but because they are different in character we should not infer the relationship *tail* PART_OF *pack* (Miller, 1998a, §1.5). MEMBER_OF is treated as a distinct relationship called *instantiation* by Lehmann (1992) and as an NT (narrower term) by Bean and Green (2001, p. 43).

The most basic relationship which is transitive and symmetric is equality between real numbers, since for real numbers $a = b$ and $b = c$ imply that $a = c$, and it is this property that has enabled untold millions of grammar school and high school students to solve algebra problems by working from one line to the next with an equals sign at the beginning of each line. The more general mathematical relationship $A \cong B$ ('A is isomorphic to B') is always transitive, roughly corresponding in language to the relationship given by the statement 'A and B are indistinguishable,' which is also symmetric and transitive.

There are many simple day-to-day relationships which are symmetric but are *not* transitive. If a and b are siblings and b and c are siblings, it follows that a and c are also siblings. But replace *siblings* with *cousins* and this ceases to be true — your cousins on your mother's side and your cousins on your father's side will not normally be cousins of one another (though this is sometimes insinuated of particularly rural areas and members of the nobility). Transitivity does not hold for the 'cousin' relationship because people are in more than one family tree. A more extreme example where symmetry *precludes* transitivity is given in electromagnetism, where opposites attract one another. If a is a positively charged particle, then a attracts every negatively charged particle, and repels every other positively charged particle. So if a attracts b it follows that b is negatively charged, and if b attracts c it now follows that c is positively charged, and so a and c repel one another. The same argument holds for north and south poles of magnets — if two poles a and c are both attracted to the pole b, you can infer that a and c have the same polarity and will repel one another.

Some antisymmetric relationships are transitive: the most basic mathematical example being that $a > b$ (a is greater than b) and $b > c$ imply that $a > c$. In day-to-day language, it normally follows that if A is above B and B is above C, than A is above C as well. These two examples are very closely related: a mathematical version of the statement

$$A \text{ is above } B$$

is expressed by the formula

$$\text{height}(A) > \text{height}(B),$$

where the *height* of an object is its coordinate (or range of coordinates) along some vertical axis (the z-axis in Figure 4.7). Now, if $\text{height}(A) > \text{height}(B)$ and $\text{height}(B) > \text{height}(C)$, then the transitivity of the relationship $>$ tells us that $\text{height}(A) > \text{height}(C)$ also, which is precisely our mathematical version of the statement "A is above C." In this way, the transitivity of the relationship 'is above' can be deduced from the transitivity of the $>$ relationship.

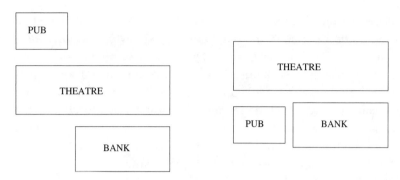

FIGURE 4.8 Is the bank next to the pub if they are both next to the theatre?

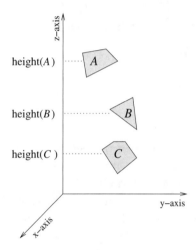

FIGURE 4.7 The transitivity of the relationship 'is above'.

We have seen that the hyponymy relationship $a \sqsubseteq b$ is transitive in almost all cases — it is certainly difficult to think of reasonable exceptions that we would actually consider to be proper hyponymy. However, there appear to be many cases where the meronymy (PART_OF) relationship is *not* transitive — to some extent, this motivates a closer look at the nature of this relationship and its definition (see box on page 115).

Spatial relationships that are largely symmetric, such as 'is near to' or 'is beside', are *sometimes* transitive, behaving more like the cousin than the sibling relationship. In Figure 4.8, the pub and the bank are both next to the theatre — whether they are next to one another as well depends on whether they are on the *same side* of the theatre.

Why such detail on such mundane points? Because these are precisely the considerations we have to think about when we try to use the graph model of Chapter 2 for learning *new* information. One fundamental objection to learning about natural language from text is that, however big a corpus we have, there are always going to be countless relationships that we *don't* see. It could be perfectly true that *plum* and *prune* are similar, and that both can be *grown, picked*, or *eaten*, but unless there is direct corpus evidence for these relationships, our model will not contain them unless we can squeeze them out some other way. This problem is referred to as *data sparseness*, and an important part of statistics and machine learning is finding ways to obviate the problems of data sparseness — our observations only ever represent a sample of the total space of all possibilities. When we know that there is going to be a shortfall — that there is guaranteed to be important information not explicitly contained in our data — then new techniques such as *smoothing* and *clustering* are needed to fill this gap (Manning and Schütze, 1999, Ch 14). Transitivity can be a powerful assumption in this process, because it allows us to infer new relationships from old ones.

4.5 Transitivity, Ambiguity and Word-Learning

Since transitivity can be a very good way of inferring information we haven't seen explicitly, it makes sense to ask whether the relationships $a \leftrightarrow b$ in the graph model of Chapter 2 should be thought of as transitive, and this turns out to have considerable practical significance when engineering language-learning systems. The answer (not surprisingly) turns out to be that treating the links in the graph as if they are transitive can yield accurate new information, but can lead to mistakes, particularly when ambiguous words are present. This leads to some interesting conclusions about ambiguous words and the way they affect measures of similarity and distance.

First of all, a few observations serve to point out how different humans and computers are when faced with ambiguity. Any time a computer needs to store a new piece of information, we assign this information to its own piece of memory with a unique name, and from then on this name cannot be used to refer to any other piece of information. For example, if I sign up for an account with a new e-mail provider or on-line retailer, if someone else has previously chosen the username `dominic`, then as far as the computer is concerned, the meaning of `dominic` has already been assigned and I had better think of a different username. How different from people! Far from insisting that a new concept should have a new name, we deliberately borrow names from the past for new individuals, products, artifacts, and

MEASURING SIMILARITY AND DISTANCE / 119

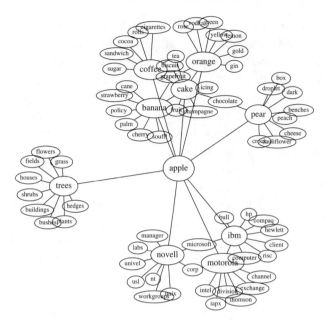

FIGURE 4.9 The words related to *apple* separate into very different clusters

cities, doing our best to leverage words and concepts which are already available, knowing that good analogies can catch people's imagination and attention, or confer positive qualities and associations whether justified or not. For example, subsequent Roman Emperors deliberately took the names *Caesar* and *Augustus*, to associate themselves with these deified heroes, and even the Russian *Czars* and German *Kaisers* named themselves after the Roman conqueror. Modern companies and institutions guard their names jealously (even names originally borrowed from common nouns and placenames), sometimes with fierce litigation, because they are worried that others will borrow their names and deliberately profit from misleading ambiguity

Given that people naturally capitalize upon ambiguity, we had better find some ways of helping computers to cope if they are to ever to understand human language, and to learn new information from old. Consider, for example, the words related to *apple* in our graph model (Figure 4.9), a typical case of a word for an everyday object which a company has subsequently chosen as its name and logo (perhaps because apples are crisp, green and healthy — they would hardly have chosen the names *peach* or *meringue* for a computer).

Suppose we want to learn *new* relationships from the graph in Figure 4.9. For example, the graph doesn't give the relationship *banana* ↔ *pear* explicitly, simply because these words didn't appear in the BNC in a coordination pattern. But we might still infer that *banana* ↔ *pear* is a good relationship, because both words are linked to *apple*. By transitivity, we can sometimes correctly infer that two words are similar in meaning even when this information was not contained in any single piece of the corpus.

Encouraged by this success, suppose we try to extend the program to other words in Figure 4.9. Not surprisingly, we soon make serious errors when we infer that *trees* ↔ *ibm* and *cake* ↔ *novell*. This is just like the situation with the buildings in Figure 4.8, where $a \leftrightarrow b$ and $b \leftrightarrow c$ implies that $a \leftrightarrow c$ only when a and c are on the same *side* of b. In the graph, we should only infer that $a \leftrightarrow c$ when the two words are both related to b *in the same way*. This problem is considered by Tversky (1977), who points out (p. 329):

> Jamaica is similar to Cuba (because of geographical proximity); Cuba is similar to Russia (because of their political affinity); but Jamaica and Russia are not similar at all.

Inferring a similarity between *jamaica* and *russia* would be particularly foolish in this situation, because geographical proximity and political affinity are *separable* features (Tversky, 1982, Gärdenfors, 2000, Ch 1). This provides some of the motivation behind the similarity measure of Hirst and St-Onge (1998): if the path in between two nodes changes direction, we may be worried that this is likely to bring a change in the nature of the similarity being inferred, and we should guard against this.

This problem is particularly acute when the node in the middle of the path is ambiguous. If a is related to b and b is related to c, but b is ambiguous, a and c are only likely to be related if they are both related to the same *sense* of b. For example, *banana* and *pear* are both related to the *fruit* (or *food*) sense of apple, and using transitivity to assume that they are also similar to one another is a good assumption in this case. On the other hand, *cake* and *ibm* are related to very different senses of *apple* (*food* and *computer company*), which is why transitivity is such a bad assumption in this case.

4.5.1 Lexical acquisition and transitivity

The task of learning information about the meanings and relationships between words from corpora, and even to learn whole classes of similar words, is sometimes referred to as *automatic lexical acquisition*, and many methods for automatic lexical acquisition use transitivity at some stage in their reasoning. Two early successes were achieved by Riloff and Shepherd (1997) and Roark and Charniak (1998), using coordination patterns to link similar nouns or entire noun-phrases.[9] Roark and Charniak describe such algorithms for extracting classes of similar words using the notion of semantic similarity, as follows:

1. For a given category, choose a small set of exemplars (or 'seed words')
2. Count co-occurrence of words and seed words within a corpus
3. Use a figure of merit based upon these counts to select new seed words
4. Return to step 2 and iterate n times
5. Use a figure of merit to rank words for category membership and output a ranked list

The figure of merit in Step 3 is effectively provided by some similarity or distance measure. The use of transitivity is in Steps 3 and 4 of this process: words similar to the old seed words are used as new seed words, and so the *next* round of similar words will be obtained by transitivity.

One problem using transitivity for this step is the danger of *infections* — once any incorrect or out-of-category word has been admitted, the neighbours of this word are also likely to be admitted. The result is that adding one mistaken entry to a class of supposedly similar objects can result in many more mistaken entries being added in its wake (Roark and Charniak, 1998, §3). These are precisely the cases where reasoning by transitivity takes us off in the wrong direction.

One successful way to avoid these infections is based on the idea that sharing *many links* in the graph is more important than having a single strong link: so reasoning by transitivity is more acceptable if it is corroborated by many different pieces of evidence.

---

[9] A noun-phrase is a grammatical unit consisting of a noun or (more general nominal) and possibly some determinative such as an article, examples including "the big bad wolf" and "some people" (see Huddleston and Pullum, 2002, p. 22).

TABLE 4.1 **Incremental clustering algorithm.**
(Widdows and Dorow, 2002, Definition 1)

For a node u in a graph G, let $N(u)$ denote the set of neighbours of u. The neighbours of a set of nodes U is the set of nodes $N(U)$ that are linked to any of the nodes $u \in U$, i.e. $N(U) = \bigcup_{u \in U} N(U)$.

The process which adds the 'most similar neighbour' to a set of nodes $A \subset G$ is as follows.

• The *candidates* $C \subset G$ for being added to the set A are those nodes which are neighbours of A but are not themselves in A, i.e. $C = N(A) \setminus A$.

• Each candidate $c \in C$ is measured according to how many of its neighbours $N(c)$ are also members or neighbours of the original cluster A.

• The best candidate $b \in C$ is the node for which this ratio is a maximum, i.e.

$$b = \mathrm{argmax}_{c \in C} \left(\frac{|N(c) \cap (N(A) \cup A)|}{|N(c)|} \right).$$

• In the case of a tie for first place, all of the winning nodes are added.

For example, the coordination "chalk and cheese" occurred many times in the BNC, making the link between these two nodes very strong (Figure 2.6). However, because *cheese* isn't related to the other neighbours of *chalk*, and *chalk* isn't related to the other neighbours of *cheese*, they should not be placed in the same cluster. At the other extreme, the top five neighbours of *sandstone* include *chalk*, *limestone* and *sand*, and the top five neighbours of *limestone* include *chalk*, *sandstone* and *sand*, so there is considerable overlap between the neighbours of these words and the neighbours of *chalk*. It follows that *chalk* should be placed in a class with *sandstone* and *limestone*, and not with *cheese*, however often the phrase "chalk and cheese" occurred in the corpus.[10]

Based on these intuitions, we developed an algorithm that adds new nodes to a cluster based on the proportion of its neighbours which are also neighbours of previous cluster members (see Table 4.1). We tested the algorithm by trying to build semantic classes (such as *tools* and *musical instruments*) based on a single initial example (such as *screwdriver* and *piano*), and the classes obtained are shown in Table 4.2.

---

[10] Another clue is that the phrase "chalk and cheese" occurs 18 times in the BNC, whereas the alternative "cheese and chalk" doesn't occur at all. Asymmetries of this kind might also be reliable clues for distinguishing idiomatic phrases from those arising from genuine semantic similarity: this conjecture remains to be tested.

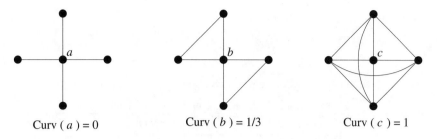

Curv (*a*) = 0   Curv (*b*) = 1/3   Curv (*c*) = 1

FIGURE 4.10 Nodes with different curvatures, depending on how interlinked the nodes around them are.

To get a reliable benchmark, we compared our results with those given by WordNet, and found that our agreement was over 80% — a big step forward over previous results. Judging by the results in Table 4.2, the success of our algorithm at reducing infection was one of the key factors: the only class that suffers badly is the *tools* class, where the ambiguity of *nail* (between the *hardware* and *body part* senses) seems to have introduced infection.

In this case, an algorithm which used transitivity, but only when backed up by several pieces of evidence, performed very well. In other words, the shortest path in our graph may be a bad measure of distance because this path may go through an ambiguous word or along an idiomatic link: however, an algorithm which infers semantic similarity from the combined evidence of *several* short paths may be much more reliable. A working version of this algorithm is available on-line,[11] showing the way the clusters are built up step by step. You should be able to test it for yourself, build a few classes of words, and see if you agree with the claims we've made.

4.5.2 Transitivity and Curvature

As well as the combined evidence of several paths being better than just one strong path, it is also clear that some words or nodes in our graph are much better for tracing paths through than others. A promising way to measure this property is to use the notion of *graph curvature*, proposed recently by Eckmann and Moses (2002) who used it to find shared interest communities on the internet.

Curvature is measured by counting the number of triangles which

---

[11] infomap.stanford.edu/graphs/build_cluster.html

TABLE 4.2 Classes of words given by incremental clustering in the graph model.

Class	Seed Word	Neighbours Produced by Graph Model
crimes	murder	*crime* theft arson *importuning* incest fraud larceny parricide burglary vandalism *indecency violence offences* abuse *brigandage* manslaughter *pillage* rape robbery assault *lewdness*
places	park	path village lane *viewfield* church square road avenue garden castle **wynd** garage house chapel drive crescent home place cathedral street
tools	screwdriver	chisel *naville* nail *shoulder* knife drill *matchstick morgenthau* gizmo *hand knee elbow* mallet penknife *gallie leg arm* sickle bolster hammer
vehicle, conveyance	train	tram car *driver passengers* coach lorry truck aeroplane *coons* plane trailer boat taxi *pedestrians* vans vehicles jeep bus buses helicopter
musical instruments	piano	fortepiano *orchestra* marimba *clarsach* violin *cizek* viola oboe flute horn bassoon *culbone* mandolin clarinet *equiluz* contrabass saxophone guitar cello
clothes	shirt	*chapeaubras* cardigan trousers breeches skirt jeans boots *pair* shoes blouse dress hat waistcoat jumper sweater coat cravat tie leggings
diseases	typhoid	malaria aids polio cancer *disease* atelectasis *illnesses* cholera hiv *deaths* diphtheria *infections* hepatitis tuberculosis cirrhosis diptheria bronchitis pneumonia measles dysentery
body parts	stomach	head hips thighs neck shoulders chest back eyes toes breasts knees feet face belly buttocks *haws* ankles waist legs
academic subjects	physics	astrophysics philosophy humanities art religion science politics astronomy sociology chemistry history theology economics literature maths anthropology *culture* mathematics geography **planetology**
foodstuffs	cake	macaroons confectioneries cream rolls sandwiches croissant buns scones cheese biscuit drinks pastries tea danish butter lemonade bread chocolate coffee milk

The words in *italics* were in WordNet (version 1.6) but not entered as members of the classes in question, and were therefore marked incorrect, which is quite strict in some cases. The words in **bold type** were correct class members which were absent from WordNet.

a particular node is part of, and comparing this with the maximum *possible* number of triangles which could include this node as a corner point. This is illustrated in Figure 4.10. The nodes a, b and c all have 4 neighbours, and if all of these neighbours are also linked to one another, there will be a total of 6 triangles (as is the case around node c). The neighbours of node b are less interwoven, producing just 2 triangles, and the neighbours of node a are totally disconnected, resulting in no triangles at all.

Definition 9 The *curvature* $\mathrm{Curv}(a)$ of a node a is the number of triangles which contain a, divided by the maximum possible number of triangles which would contain a if all the neighbours of a were connected to one another.

If a node a has n neighbours (in which case we say that n is the *degree* of a), the total number of triangles which a could be part of is equal to $n(n-1)/2$ (every pair of neighbours p and q makes a triangle with a if there is a link $p \leftrightarrow q$, therefore the number of triangles is equal to the number of possible pairings, which is $n(n-1)/2$, a nice exercise to prove). If t is the number of actual triangles which include a as a corner, the curvature will therefore be given by the fraction

$$\mathrm{Curv}(a) = \frac{2t}{n(n-1)}.$$

For the examples in Figure 4.10, the degree of the central node is 4, so the number of possible triangles is 6, and the respective curvatures are

$$\mathrm{Curv}(a) = \frac{0}{6} = 0 \qquad \mathrm{Curv}(b) = \frac{2}{6} = \frac{1}{3} \qquad \mathrm{Curv}(a) = \frac{6}{6} = 1.$$

For two nodes p and q drawn at random from the neighbours of a (two nodes such that $a \leftrightarrow p$ and $a \leftrightarrow q$), the curvature of a gives the probability that there is also a link $p \leftrightarrow q$. The curvature of a node a is thus a precise measure of how often the transitivity assumption is valid for its immediate neighbours.

In general, words with *high* curvature scores are often very informative and unambiguous: of the frequently occurring words represented in our graph, the overwhelming majority of high-curvature words are names of *countries*. At the other extreme, many of the words with the *lowest* curvature are names of *people*. Names of countries are generally unambiguous (with some exceptions), and names of people are highly (and deliberately) ambiguous. This led to the hypothesis that there is a strong negative correlation between curvature and ambiguity, which has been borne out by our experiments.

Graph curvature is an interesting concept, partly because curvature is traditionally associated with *continuous* models and our graph is a *discrete* model. In the next chapter, we will focus on using continuous models for discrete data (in the building of vector models), and curvature is a good preparatory example.

4.6 Splitting the Semantic Atom

Since the transitivity assumption is unjustified for ambiguous words, causing problems such as infection for lexical acquisition, is it possible to run this argument the other way and use the breakdown of transitivity to *recognize* which words are ambiguous?

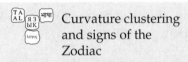

Curvature clustering and signs of the Zodiac

Among the less frequent words in our graph, the curvature measure sometimes does a remarkable job at assembling clusters, by taking connected regions of the graph all of whose nodes have curvature above a given threshold. One empirical example is the following cluster containing 10 of the 12 signs of the Zodiac:

Aries Taurus Gemini Virgo Libra Scorpio Sagittarius Capricorn Aquarius Pisces

This is extremely accurate for such an empirical method. There are no mistaken interlopers, and the only two signs missing are *Cancer* and *Leo*, which have lower curvature because they are both ambiguous.

Such a hope is motivated by graphs such as the one of *apple* in Figure 4.9, and the graphs of the words *arms* and *head* we saw in Figures 2.8 and 2.9. Humans have no trouble in looking at such pictures and seeing that different groups of words correspond to different meanings of the word *apple*. Is it possible to formalize this intuitive grasp so that computers can do the same?

In Figure 4.9, *apple* is the only node which connects the top cluster of foodstuffs with the bottom cluster of computer companies. If it were not for the *apple* node, there would be no 2-step path from (say) *banana* to *ibm* or from *cake* to *motorola*. It follows that *removing* the *apple* node would lead the graph to fragment into disconnected regions.

A simplified version of this situation is shown in Figure 4.11 (which is like the graph around the node b in Figure 4.10). The *apple* node is torn between two triangles, and this suggests the hypothesis that the word represented by *apple* should really be thought of as two senses *apple*$_1$ and *apple*$_2$, belonging to the upper and lower triangles respectively. Note that the graphs of *arms* (Figure 2.8) and *head* (Figure 2.9) also contain subgraphs which have the butterfly or bowtie shape of Figure 4.11.

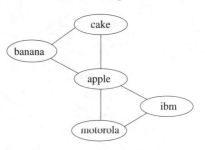

FIGURE 4.11 Neighbours of *apple* in different triangles.

This suggested the following criterion for recognizing ambiguous words and their different senses:

Conjecture 1 (Widdows and Dorow, 2002, §6)
A node in the graph represents an ambiguous word if its removal causes the graph around that word to fragment into disconnected regions.

In this case, each disconnected region corresponds to a different *sense* of the word.

This intuitive conjecture has worked in many cases. Another elegant example, where an old word has been borrowed in recent years to describe a piece of computing equipment, is given by the graph around the word *mouse* in Figure 4.12.[12] If we remove the *mouse* node from the centre of this graph, we are left with two separate regions, corresponding exactly to the little grey rodent (bottom right) and the computer mouse (top left).

---

[12]This figure was produced by Beate Dorow (Dorow and Widdows, 2003) using a database of relations extracted from text by Dekang Lin (Lin and Pantel, 2002). Lin's data has separate entries for multiword expressions like *landing gear* and *hydraulic system*, so the preprocessing of the corpus is more sophisticated than that used for our BNC model in this chapter.

128 / GEOMETRY AND MEANING

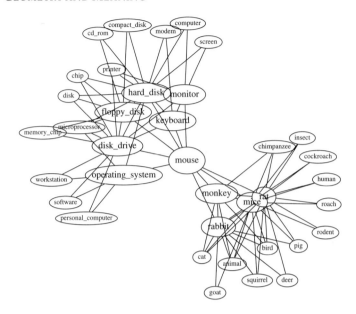

FIGURE 4.12 The neighbours of *mouse* decompose exactly into regions corresponding to two different senses.

One intuitive analogy I have often used, which clearly links the properties of ambiguous words to the way distances are measured, is to say that an ambiguous word in our graph behaves like a 'semantic wormhole.' If you enter the *mouse* node along the *rat*↔*mouse* link, and leave along the *mouse*↔*monitor* link, then the ambiguity of *mouse* has effectively teleported you from one region of semantic space (animals) to another (computing equipment).

Needless to say, things often aren't so simple, partly because our data is never perfect, but more importantly because many ambiguous words are more subtle than this. We have so far assumed that the different senses of an ambiguous word are semantically unrelated to one another, so that removing this particular word removes any association between the different meanings. Consider, by contrast, the diagram of words related to *wing* in Figure 4.13. There are many different meanings of *wing* represented in this graph — the wing of a bird, the wing of an aircraft, the wing of a building and the wing of a political party. Rather than being unrelated concepts for which we happen to use the same word, all of these senses are *systematically related*.

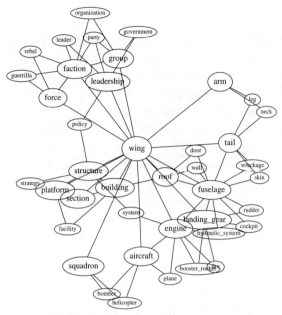

FIGURE 4.13 Words related to different senses of *wing*

There are often many systematic relationships between different senses of the same word — more are evident in the graph of *apple* in Figure 4.9, since an *apple* can be a *fruit*, a *tree* and a *foodstuff*. Systematic ambiguity is investigated by many authors including Pustejovsky (1995) and Buitelaar (1998), and while there is much work still to be done, our understanding of this phenomenon in English and other languages is improving.

The goal of 'splitting the semantic atom' is ambitious, and the reader should not be misled into believing that this work has been successfully completed. There is a great deal of work still to be done in discerning appropriate units of meaning and updating these units when old words are used in new ways, a habit that is ubiquitous in our language (especially in the marketing age). However, the examples in this section show that we have made significant progress, at least with the simple cases where a single ambiguous word is a combination of two very different senses.

Conclusion

This chapter has introduced some of the standard systems for measuring distances, casting some light (and some doubt) on their effectiveness in modelling distances between words and meanings. The effect of the mathematical assumptions underlying different methods has a pervasive underlying influence which can become overlooked in implementations, especially when we have complicated and subtle linguistic phenomena such as ambiguity to consider. One of the messages of this chapter is that we should be careful when adapting mathematical techniques to new areas, lest we reason that a computer is like a banana because they are both related to apple (Figure 4.9).

Wider Reading

A readable and much more complete introduction to metrics can be found in Gärdenfors (2000, Ch 1), which is definitely recommended to those who are less familiar with this branch of mathematics. This chapter also discusses the complementary roles played by similarity and distance measures, and some of the transformations between them that have been used in psychology. Those looking for a more thorough mathematical account of the properties of different metrics should consult Hausdorff (1914, Ch 6), which, though original, hardly appears dated at all, and puts metric spaces in the context of point sets and algebra — eager young mathematicians will find themselves delving into the rest of the book before they know it.

Some of the properties of words expressing spatial relationships, described briefly in this chapter, are discussed by Herskovits (1986). In particular, Chapter 3 of this book analyzes some of the uses of spatial relationships such as "is near to," finding that they are not as symmetric as may be supposed, but that the ordering of objects can be used to express a Figure–Ground distinction (not unlike the Prototype–Variant distinction). For example, it would be considered more normal to say "The parked car is near to the Empire State Building" than to say "The Empire State Building is near to the parked car." The relationship between these asymmetries and fundamental case-roles in linguistics is highlighted by Delancey (2000).

The idea of generating classes of words by finding words which are similar to a particularly central 'seed-word' is very similar in spirit to the *prototype theory* founded by Eleanor Rosch (1975) and collaborators such as Carolyn Mervis (Mervis and Rosch, 1981). The idea behind prototype theory is that people classify objects (such as birds) based on their similarity with a prototypical bird (such as a robin). Good experimental evidence for prototype theory comes from the fact that there is a remarkable degree of agreement among subjects as to which objects are better examples of their categories than others. Good secondary introductions to prototype theory and its contribution to conceptual models can be found in Gärdenfors (2000) and Aitchison (2002, Ch 5).

The effect of ambiguity on the structure of WordNet, seen as a graph, is described by Sigman and Cecchi (2002), who also note that ambiguous words contribute greatly to the connectivity of the lexicon, linking areas that would otherwise be distant. The idea of finding the different senses of an ambiguous word w by dividing (or amalgamating) the neighbours of w into different classes or *clusters* is discussed further in Section 6.3. Within the graph model itself, we have used *Markov Clustering* (van Dongen, 2000) to find different meanings of words in a graph model (Dorow and Widdows, 2003).

5

Word-Vectors and Search Engines

So far in this book we have discussed symmetric and antisymmetric relationships between particular words in a graph or a hierarchy, described one way to learn symmetric relationships from text, and shown how to use ideas such as similarity measures and transitivity to find nearest neighbours of a particular word. But ideally we should be able to measure the similarity or distance between *any* pair of words or concepts. To some extent, this is possible in graphs and taxonomies by finding the lengths of paths between concepts, but there are at least three problems with this. First of all, finding shortest paths is often computationally expensive and may take a long time. Secondly, we may not have a reliable taxonomy, and as we've seen already, the fact that there is a short path between two words in a graph doesn't necessarily mean that they're very similar, because the links in this short path may have arisen from very different contexts. Thirdly, the meanings of words we encounter in documents and corpora may be very different from those given by a general taxonomy such as WordNet — for example, WordNet 2.0 only gives the *fruit* and *tree* meanings for the word *apple*, which is a decided contrast with the top 10 pages currently returned by the Google search engine when doing an internet search with the query *apple*, which are all about Apple Computers.

Another limitation of our methods so far is that we have focussed our attention purely on individual concepts, mainly single words. Ideally, we should be able to find the similarity between two *collections* of words, and quickly. For this, we need some process for *semantic composition* — working out how to represent the meanings of a sentence or document based on the meaning of the words it contains. This all sounds like a pretty tall order, but (to some extent) it's actually been possible for years and you will probably have encountered such a system many times — it's precisely what a search engine does.

The traditional ways in which a search engine accomplishes these tasks are almost entirely mathematical rather than linguistic, and are based upon numbers rather than meanings. Using its distribution in a corpus, the meaning of each word is represented by a characteristic list of numbers, and the numbers representing a whole document are then given simply by averaging the numbers for the words in the document. Bizarre but effective!

This chapter is all about the lists of numbers used to represent words and documents, and these lists are called *vectors*. The discussion of vectors will finally bring us to talk in detail about different *dimensions* and what they mean to a mathematician. This in itself might be of interest to many readers, and quite different from what you supposed.

5.1 What are Vectors

A *vector* is a very useful way of keeping track of several different pieces of information, all of which relate to the same concept or object. For example, suppose I have a drawer at home where I keep my leftover currency from travelling to different countries, and have a small pile of 6 US dollars ($6), 20 UK pounds (£20), 15 Euros from Germany (€15) and 100 Japanese yen (¥100). We could write this information down in a table

$	£	€	¥
6	20	15	100

or if we keep careful track of the convention ($, £, €, ¥) we could write this as the *row vector*

$$Dm = (6, 20, 15, 100),$$

where Dm stands for "Dominic's money." Now, suppose that Maryl has a similar drawer, which contains

$	£	€	¥
50	16	20	0

This information can be encoded in the row vector

$$Mm = (50, 16, 20, 0).$$

If we marry our fortunes together, what will our combined currency drawer contain? The answer is simple — just add together the numbers in the matching positions and you get the combined vector

$$Dm + Mm = (56, 36, 35, 100).$$

Suppose that we decided to put this money into savings and it grew by 20% (which corresponds to multiplying by the number 1.2). Then we'd have

$$1.2(Dm + Mm) = (67.2, 43.2, 42, 120).$$

Not impressed? Maybe it doesn't seem like rocket science to write down two lists of numbers, keep track of which numbers refer to which currencies, and then add and multiply these numbers. But it turns out that the ability to break a situation down into individual numbers, to do separate calculations with those numbers, and then to combine the answers to represent a new situation, can be extremely powerful, because it allows us to break down a potentially complicated process into a number of extremely simple ones. And by the way, you just dealt with points in a *four-dimensional* space without even blinking.

5.2 Journeys in the plane

Another situation that can be described using vectors is one we've already encountered — namely, the 2-dimensional plane can be thought of as a vector space. In Figure 5.1, the arrows a and b represent two 'journeys' in a plane. The combined journey from going along the a arrow and then the b arrow is called the *sum* of the journeys a and b, and so is (naturally enough) written as $a + b$.

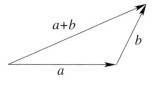

FIGURE 5.1 The journey $a + b$

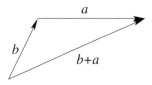

FIGURE 5.2 The journey $b + a$

On the other hand, we could have gone along the journey b first, followed by the journey a, and this combined journey would be called $b + a$. One fundamental property of the flat plane is that these two journeys have the same destination: $a + b$ and $b + a$ are two ways of writing the same journey. This isn't always true — it depends on the space in which a and b are journeys. For example, imagine your space is a sphere like the earth and you start on the equator. If you go 1000 miles north and then 1000 miles east, you

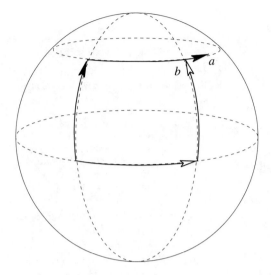

FIGURE 5.3 When combining two journeys on a sphere, you can end up at different places depending on which journey you make first.

will end up further east than if you go 1000 miles east, *then* 1000 miles north, because the parallel (line of latitude) 1000 miles north is a smaller circle than the parallel of the equator, as shown in Figure 5.3.[1] Because a sphere is curved, it turns out that the order in which you make different journeys matters — the convenient identity $a + b = b + a$ ceases to be true.

In fact, the amount to which different journeys can land you in different destinations can be used to define and measure the very concept of the *curvature* of a mathematical space. Vector spaces are special precisely because they *aren't* curved — they are flat or 'linear' (mathematicians often say *Euclidean*), and because of this the study of vectors is called *linear algebra*. As a result of this linearity, vector spaces obey Euclid's fifth axiom of geometry, which states that parallel lines never meet.

---

[1] At the extreme, if you go as far as the north pole, the line of latitude collapses from a circle to a single point — the north and south poles do not have a well defined longitude, since *all* lines of longitude intersect at the poles. This singularity provides the solution to the brainteaser

> Suppose you walk ten miles south, ten miles east, ten miles north and find yourself at the point you started from. Where are you?

the standard solution being that you are at the north pole. Other solutions lie just north of the south pole.

The process of putting two journeys, or vectors, together into a single journey is called *vector addition*. You should convince yourself that this is a wise generalization of the real-number addition operation you learnt as a child. In this context, real numbers behave as 1-dimensional vectors. Draw yourself a mental picture of how our $a + b$ and $b + a$ journeys appear along a single line, and I hope you'll see what I mean. Another way of describing vector addition is called the parallelogram rule. If you can see why by mentally combining Figures 5.1 and 5.2, you're well on the way to understanding what's going on.

The symbol $+$ which means "add together these two vectors (or numbers)" is called an *operator*. An operator is different from a relationship such as similarity $\leftrightarrow$ or hyponymy $\sqsubseteq$, because whereas $a \leftrightarrow b$ is just a statement that the relationship

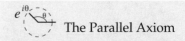 The Parallel Axiom

Euclid's fifth *Axiom of Geometry* states

5. That, if a straight line falling on two straight lines makes the interior angles on the same side less than two right angles, the two straight lines, if produced indefinitely, meet on that side on which the angles are less than the two right angles.

This is equivalent to saying that parallel lines never meet. Because it is so much more cumbersome than the first four axioms, mathematicians tried to prove it as a consequence of these, rather than assuming it as an axiom in its own right, for over two thousand years.

In the 19th century, mathematicians such as Gauss, Bolyai, Labachevsky and Riemann finally realized that the behaviour of parallel lines depends on the nature of the space you are working in: for example, all the lines of longitude on a globe are parallel at the equator and meet at the poles. Letting go of the parallel axiom led to a whole new field of *non-Euclidean geometry*, which paved the way for the Theory of Relativity.

holds, the operation $a + b$ has a result or an outcome. Familiar examples of relationships between numbers are equals ($=$) and less than ($<$). Familiar examples of operators are multiplication ($\times$), addition ($+$) and subtraction ($-$). The similarity and distance measures of Chapter 4 are all operators because they produce a number as their outcome.

Just as a relationship $a \sim b$ where the order of a and b is interchangeable is called symmetric (Definition 2), an *operator* for which the order of the inputs is interchangeable also has a special name.

Definition 10 A mathematical operator ∘ is said to be *commutative*[2] if $a \circ b = b \circ a$ for all possible a and b.

Simple examples of commutative operations are addition and multiplication of real numbers: we know that $a + b = b + a$ and $ab = ba$. Subtraction, on the other hand, is certainly *not* commutative, because $a - b \neq b - a$. In fact, since in general

$$a - b = -(b - a),$$

the result of swapping a and b round is to give us the *opposite* answer, and for this reason the subtraction operator is *anticommutative*.

The operation of adding the currency vectors in Section 5.1 is also commutative — it doesn't matter whether you add my fortune to Maryl's, or the other way round, you get the same answer. Vector addition must always be commutative, and this is one of the reasons why vector addition is a good generalization of real number addition. In fact, the operation of amalgamating of our currency vectors is commutative precisely because it is a combination of four separate addition operations with numbers, and each of these separate sums is commutative.

So much for adding vectors. The other operation we must be able to perform is *scalar multiplication* — stretching or shrinking of vectors. Any of our journey vectors can be scaled up or down to give you a journey in the same direction but of different length. Similarly, we could scale our currency vector by a factor of 20% (multiplying by 1.2), or any other factor (though in this case we might find ourselves rounding the result to two decimal places or the nearest cent, which the bank normally does on our behalf when calculating interest). Just as addition must satisfy a few properties such as being commutative, scalar multiplication must also be well-behaved in certain ways. For example, if Maryl and I invested our money separately, each got a return of 20%, and then married our new fortunes together, we should get exactly the same amount as if

---

[2] The word *commutative* has nothing to do with travelling to work or having a prison sentence reduced: nor is there really any good reason why we couldn't call an operator *symmetric* instead of coining this new technical term. There is no particularly good reason why you have to learn a new four syllable word at this point, except that for those readers who are already familiar with it, it would be too confusing to change things now. (Because of this, the excessive jargonization of all contemporary fields is getting out of hand and is apparently impossible to resist, the proliferation of science itself scattering us in a technological Tower of Babel.)

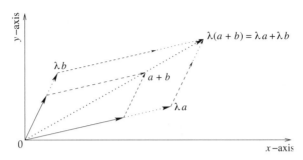

FIGURE 5.4 Adding two vectors together and then rescaling gets you to the same place as rescaling both vectors first and adding the results. This is another characteristic of Euclidean (flat) space.

we married them together first and then increased the total by 20%. In equations, this would be

$$1.2Dm + 1.2Mm = 1.2(Dm + Mm),$$

which you can easily check is true in this case. An illustration of this property is given in Figure 5.4. This scaling operation can be traced back to Euclid's second Axiom of Geometry, which states that it is always possible

   2. To extend a finite straight line continuously in a straight line.

A set of points equipped with addition and scalar multiplication operations must satisfy the eight axioms of Figure 5.5 to be a *vector space*. These axioms were given by Peano in 1888, though this was really a culmination of different strands of mathematical progress throughout the 19th century and before. In the next section we will explain how this culmination came to be, and this process will hopefully make some difficult ideas very clear. This will leave us in a sound position to use the techniques of vector spaces to describe words and their meanings.

5.3 Coordinates, bases and dimensions

What do the currency vectors of Section 5.1 and the journey vectors of Section 5.2 have in common? We could try to go through and check that the eight axioms of Figure 5.5 hold for both systems and say that this *is* what they have in common, but this is really a topic for a linear algebra homework assignment, not a book on meaning. What we want to know is how they came to be regarded as similar systems — after

> **Formal Definition of a Vector Space**
>
> A (real) Vector Space is a set V equipped with two mappings, called addition (which maps $V \times V$ to V) and scalar multiplication (which maps $\mathbb{R} \times V$ to V).
>
> Addition must obey the following axioms:
> - A1. Addition is associative, so that for all $a, b, c \in V$, $(a + b) + c = a + (b + c)$.
> - A2. Addition is commutative, so that for all $a, b \in V$, $a + b = b + a$.
> - A3. There is an additive identity element $0 \in V$ (called the "zero vector") such that for all $a \in V$, $a + 0 = 0 + a = a$.
> - A4. For each element $a \in V$, there exists an element $-a \in V$ with $a + (-a) = 0$.
>
> Scalar Multiplication must obey the following axioms:
> - M1. Scalar multiplication is associative, so that for all $\lambda, \mu \in \mathbb{R}$ and for all $a \in V$, $(\lambda \mu)a = \lambda(\mu a)$.
> - M2. $1a = a$ for all $a \in V$.
> - M3. Scalar multiplication is distributive over addition in V, so that for all $\lambda \in \mathbb{R}$ and for all $a, b \in V$, $\lambda(a + b) = \lambda a + \lambda b$.
> - M4. Scalar multiplication is distributive over addition in $\mathbb{R}$, so that for all $\lambda, \mu \in \mathbb{R}$ and for all $a \in V$, $(\lambda + \mu)a = \lambda a + \mu a$.

FIGURE 5.5 A vector space must satisfy these eight axioms (Jänich, 1994, p. 17)

all, this similarity was recognized long before the axioms were written down.

The main breakthrough in thinking of journeys in the plane as lists of numbers is made by giving each journey a set of *coordinates*, which break the journey down into different components in different directions. You will almost certainly have come across these in high school, and we have used them already in this book.

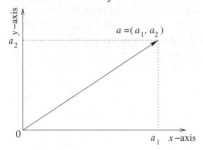

FIGURE 5.6 The coordinates of a vector

The basic picture is in Figure 5.6, where the vector a is decomposed into 2 parts by *projecting* it onto two fixed vectors or *axes*. (The word *project* comes to us through the Medieval Latin *projectum*, meaning *throw forward*.) If the projection hits the x-axis at a distance of a_1 units from the origin, this number is said to be the x-coordinate of the point a, and similarly, if the projection hits the y-axis at a_2, then this is said to be the y-coordinate of a. The point a is then said to have coordinates (a_1, a_2), though it is only by convention that the first coordinate refers to the horizontal component and the second to the vertical component, the 'right' and 'up' directions regarded as positive.[3]

There is one confusion to avoid. We've talked about giving coordinates to points in the plane and to journeys in the plane. but a journey requires a pair of points (a beginning and an end), not a single point, so how can both these ideas be represented by the same sort of coordinates? The answer is that the point a is identified with the journey *from the origin* or zero point to the point a, so we can think of both "the point a" and "the journey from 0 to a" as vectors in the plane.

Giving these numbers or coordinates to a journey vector enables us to carry out the processes of vector addition and scalar multiplication without having to draw the pictures. For example, we can work out the additions in Figures 5.1 and 5.2 very easily: if the vector a has coordinates (a_1, a_2) and the vector b has coordinates (b_1, b_2) then their sum $a + b$ will have the

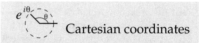

Cartesian coordinates

The use of a pair of numbers such as (x, y) to represent points in the plane is justly associated with Rene Descartes (1596-1650), though he was not the first to invent it. Descartes realized that many of the relationships between geometric figures such as lines and curves could be represented by numbers in this way — for example, the cosine wave on page 106 is the locus of all points whose Cartesian coordinates (x, y) are related by the equation $y = \cos(x)$.

In this way, many old geometric problems, such as finding the point where two curves intersected, could be solved using the techniques of algebra which, thanks mainly to Islamic mathematicians, had made enormous progress since the Greeks. In algebraic terms, the use of Cartesian coordinates amounts to representing the plane as a set of pairs of real numbers in the product set $\mathbb{R} \times \mathbb{R}$, which is why the plane is often referred to by the symbol $\mathbb{R}^2$ (normally pronounced "R2," like the cute android from *Star Wars*).

coordinates $(a_1 + b_1, a_2 + b_2)$. The method of using coordinates in this fashion was a gradual mathematical development: Appolonius of Perga (ca.260-190 BC) and the French bishop Nicole Oresme (ca.1320-1382)

---

[3]That this is only conventional is apparent in computer graphics, where the 'down' direction is normally regarded as positive and the position of a point is determined by its coordinates measured from the top left hand corner of the screen.

both described points in figures using distances from a pair of fixed locations, ideas which contributed to the *Analytic Geometry* of Descartes (1637). The description given by Descartes of choosing a particular line as a unit, and ascribing numerical lengths to other lines according to their ratio with this unit, is one of the first clear-cut definitions of 'measurement' in the sense of Section 1.4.3.

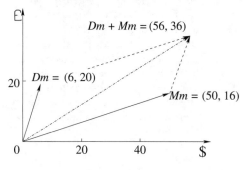

FIGURE 5.7 Adding together currency vectors

This explains how we can use numbers to represent the pictures of Figures 5.1 and 5.2. We can also go the other way — suppose we wanted to represent the first two coordinates of the currency vectors in Section 5.1 pictorially. (These are the amounts of dollars and pounds, $Dm = (6, 20)$ and $Mm = (50, 16)$.) We could plot these on a grid and add them together using the parallelogram rule, and finally read off the coordinates of the resulting sum, $Dm + Mm = (56, 36)$, as in Figure 5.7. This may seem like a long way round, though as we'll see in Chapter 6, representing vectors as points on a plane like this, rather than as a pair of numbers, can give a much more intuitive representation of which groups of vectors are close to one another. Just as points in the plane are often given x and y coordinates, Figure 5.7 shows that we can just as well talk about "$ and £ coordinates." Ideally, we would have represented the "€ and ¥ coordinates" as well, though it would be difficult to represent three of these coordinates at once on a flat page, and pretty much impossible to represent all four of them in a way that made intuitive sense.

In order to represent a vector as a list of numbers, it is clearly necessary to keep a 'key' telling us which quantity each number refers to. For the currency vectors, we needed to remember that the code (a_1, a_2, a_3, a_4) was shorthand for "$$a_1$, £$a_2$, €$a_3$ and ¥a_4." When working in the plane we need to choose coordinate axes, and while the convention is to use 'eastward' and 'northward' pointing axes, there is no *a priori* reason for making this particular choice.[4] This key telling

---

[4] We could use 'south' and 'north-northwest' as our coordinate axes, since you can

us which coordinate is which is called a *basis* for the vector space. For example, the basis for our currency vectors is the set $\{\$1, £1, €1, ¥1\}$.

Now, each coordinate refers back to one of these basis elements, so the number of coordinates needed to represent any point is always the same as the number of items in the basis. For example, the vector $Dm = (6, 20, 15, 100)$ has 4 coordinates, one to refer to each of the 4 currencies listed in the basis. This number is an important characteristic property of any vector space, and it's a word you will have come across many times.

Definition 11 The *dimension* of a vector space is the number of coordinates needed to specify a given point uniquely.

The flat plane has two dimensions, since each point needs 2 coordinates to represent it: normally these are the x and y coordinates. The currency vectors of Section 5.1 have *four* dimensions, because each vector is made up of 4 coordinates (or numbers) of Dollars, Pounds, Euros and Yen.

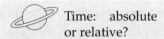 Time: absolute or relative?

The distinction between a vector itself and the coordinates used to write the vector down may be subtle, but it is a vital conceptual difference. For example, (according to the Special Theory of Relativity) the time measured between different events is not a property of the events themselves, but depends on how you choose your coordinates.

Because the speed of light c is *constant*, and speed is a relationship between position and time, if you change the way you measure position you must also change the way you measure time. The relationships between the time coordinates of different events is therefore different for different observers. This interplay between space and time has strange consequences — for example, two events which are simultaneous for one observer may happen at *different* times for another. If this sounds confusing but you really want to get it, see www.puttypeg.com/train.html.

---

still specify any journey by saying how far south you have to go, followed by how far north-northwest you have to go — and this representation would be just as unique and unambiguous as the more familiar x and y coordinates. When developing the technique of using coordinates, Descartes in fact used many different pairs of axes, not necessarily at right angles to one another — the convention of using the same pair of perpendicular axes for most problems came later.

Hopefully this will break a few misunderstandings, if they haven't been broken already. We've probably all heard and wondered about the question

> Is time the fourth dimension?

Clearly the answer for our currency vectors is "no" because in this space, "Japanese Yen" is the fourth dimension! This goes back to the difference between studying physical *space* and mathematical *spaces*. In mathematics, we use however many dimensions we need to represent the information we're working with — if a situation is complicated enough to require more than than 2, or 3, or 4 coordinates to represent it properly, then we'll use a space with as many coordinates as are necessary. However, if you *are* seeking to model the physical universe, then four coordinates can be a very good choice — three to measure different directions in space, and one to measure the time that elapses between two events (this is exactly what we did when defining the *Minkowski distance* on page 110). So in this sense, the answer is "yes," insofar as there are some very good models of the physical universe in which time is a fourth dimension.

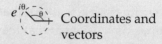

Coordinates and vectors

Since there are many different ways to choose a basis, and thus to assign coordinates to each point in a vector space, we must check that the mathematics is the same whatever choice we make. For example, if the dimension of a vector space is the number of elements in any basis, we had better make sure that this number is the same for *every* possible basis of a given vector space (Jänich, 1994, p. 46) — otherwise 'dimension' would not be a property of the space itself, but only of one particular coordinate system.

Also, the sum of two vectors $a + b$ can be calculated by adding their respective coordinates, but must be the same vector for every possible coordinate system. These important properties can all be proved from the axioms in Figure 5.5.

However, just because a fourth dimension is a useful concept, it doesn't make it an *easy* concept, and the difficulties have been noted since vectors were first invented. The term *vector* was probably first used by Sir William Rowan Hamilton in his work on vectors in three and four dimensions (which he called *quaternions*). His verbal description of how a fourth direction might be perpendicular to the North-South, East-West *and* Up-Down directions all at the same time is quite detailed and at the same time almost desperate (Hamilton, 1847). The situation is touchingly parodied in the novel *Flatland* (Abbott, 1884, Ch. 21), where a square vainly tries to coax his hexagonal grandson into the speculation that there might be a direction that is "Upward and not Northward."

The main trouble that most people have when trying to handle more than three dimensions is that they try to visualize all these dimensions at once. We already accepted this limitation in Figure 5.7: when I was explaining what currency vectors and journey vectors have in common, I only drew a picture of two of the currency coordinates (dollars and pounds) since we could easily represent these on a flat piece of paper. Do not hold yourself back from understanding what dimensions are by trying to visualize more than three dimensions at once: it's a good way of getting frustrated because it's just not within our physical experience. Spaces with many dimensions are easy to live and work with once you accept that several dimensions are perfectly valid and consistent, without trying to forcibly reconcile each dimension with visual experience simultaneously.

5.4 Search Engines and Vectors

Early search engines relied on simple Boolean matching algorithms, which will be described in Section 7.1. If a document matched the keywords in a user's query, it was marked relevant and returned to the user: if not, it was marked irrelevant and left behind. But by the 1970's, as document collections grew bigger, the flaws in such an approach became more and more damaging. One main problem that people encountered is that with large document collections (which would be considered small by today's standards), returning *every* document that matched the keywords would still leave the user with a huge mass of documents to wade through to try and work out if they were relevant. For example, suppose that you wanted to find out about helicopters. A search of the internet that returned *every* matching document might send you countless news articles describing rescue missions that involved a helicopter at some point, but these may not tell the you very much about the helicopters themselves (though they may tell you something about the contexts in which helicopters appear).

TABLE 5.1 Recording the number of times each indexed term occurs in each document

	Document 1	Document 2	Document 3
bank	0	0	4
bass	2	4	0
commercial	0	2	2
cream	2	0	0
guitar	1	0	0
fishermen	0	3	0
money	0	1	2

To take more extreme examples, a dictionary will probably contain most common words that aren't proper names, but the dictionary isn't relevant to every user whose query contains any of these words. A telephone directory will contain a huge number of proper names, but it still won't be relevant to many queries containing one of these names. Long documents containing huge numbers of words could be declared relevant to almost any query — but these are often precisely the documents whose information is most diluted. It became very clear that a better measure of relevance was needed than could be achieved by simply partitioning the document collection into documents that did and didn't contain the keywords.

One way to do this (and I should stress that there are several) is to represent words using vectors. The coordinates of these vectors are numbers which measure the extent to which a particular word is important to a particular document. To begin with, these coordinates are obtained simply by counting the number of times each word is used in each document.

As a small-scale example of this technique, consider the three documents in Figure 5.8, in which a handful of different words have been highlighted (*bank, bass, commercial, cream, guitar, fishermen* and *money*). The number of instances (tokens) of each of these words in each of the three documents is recorded in Table 5.1.

Such a table, which records the number of times each word occurred in each of a collection of documents, is called a *term-document matrix*.[5]

---

[5] A *matrix* (plural *matrices*) in this context is simply a table of numbers. There are well-defined algebraic rules for how pairs of matrices can be added and multiplied together, provided that the number of rows and columns in the two matrices is compatible.

Document 1 (BNC)
The first **Bass** VI I remember seeing was being used on television by Jack Bruce, during a mimed performance of Strange Brew by **Cream**. Recollections indeed: an extremely fashionable **Cream**, with serious sideburns all round, grossly extended collars on satin shirts, Clapton's Hendrix perm, flared trousers and Les Paul. Being nobbut a sprog, I remember wondering why bassist Bruce was playing a **guitar**. Only he wasn't, of course; he was playing a Fender **Bass** VI. He also became arguably the most famous exponent of the instrument, along with Eric Haydock of The Hollies.
Document 2 (NYT, 1996)
ORLEANS, Mass. — When weekend **fishermen** looking to unwind come to Rock Harbor, they take out their rods and reels and angle for striped **bass** one by one. When **commercial fishermen** like Mike Abdow go out on their boats to earn a living, they catch **bass** the same way. By law, they must reel them in one at time, without nets, traps or long-line gear. But right now, the waters are rough between people who fish **bass** for a living and those who angle for pleasure. In a feud that has divided people along the Massachusetts coast, big-**money** sporting interests are trying to stop small-time **commercial fishermen** from pulling in any more striped **bass**.
Document 3 (NYT, 1995)
Kuala Lumpur, Oct. 26 (Bloomberg) – **bank** Negara will change the way it calculates the cost of lending **money** to **commercial** banks and financial institutions. In a statement, the **bank** said that as of Nov. 1, the base lending rate, at which **commercial** banks can borrow from the central **bank**, will become more responsive to movements in **money**-market rates. Several weeks ago, the **bank** said it would change the way it calculates the base rate, said Desmond Ch'ng, a banking analyst OCBC Securities (Malacca) Sdn. Bhd.

FIGURE 5.8 Three documents about different topics

The snapshot we've taken in Table 5.1 is only a tiny fragment of the term-document matrix used by any real search engine — we've only taken three documents and only considered a few of the words they contain. However, it's still possible to get some idea of how the table works, and hopefully the reader will then be able to extrapolate and imagine the huge table we'd produce from a document collection of several million words or even several billion webpages.

Table 5.1 can be used as a simple inverted file index, like a traditional concordance.[6] That is, if you're interested in *fishermen*, the table will tell you to look in Document 2, if you're interested in a *bank*, the table will tell you to look in Document 3, and so on. But the real advantage of measuring the *number* of times each word occurs in a document comes when you have words occurring in many documents, and you want to know which document is the most relevant. For example, if you want to find out about *money*, Table 5.1 will tell you that it's mentioned in both Document 2 and Document 3, but since it's mentioned *twice as much* in Document 3, you should start by reading this document, because the term *money* is more concentrated or denser in this document. This would be good advice — while Document 2 mentions tensions between two groups of people which have monetary consequences, Document 3 is directly about finance.

There are many problems and issues with this idea that we haven't even begun to address, some of which are listed below:

- Many possible meanings of the different terms are absent from this fragment — *bank* can also refer to a *river bank*, and most of the time *cream* is more likely to mean a *dairy product* than a *1960's rock band*.
- The term *bass* does have more than one meaning in our 3 documents — it means a kind of *musical instrument* in Document 1 and a kind of *fish* in Document 2. If a user is only interested in one of these meanings (and in this case, it's unlikely that they'd be interested in

---

Matrices grew out of an an 1858 memoir on transformations written by the British mathematician Arthur Cayley (1821-1895), partly as a generalization of Hamilton's *quaternions* (Boyer and Merzbach, 1991, Ch. 26).

[6]The term 'inverted file index' or 'inverted file format' is used because words are described by a list of documents, whereas it's more usual to think of documents being described by a list of words (Salton and McGill, 1983, Ch. 2). A concordance, often of the Bible, is a big book where a Biblical scholar can look up a word such as *light* and find the chapter and verse of every place where the word *light* appears, hoping to trace the way a theological concept is used and developed through the centuries (Witten et al., 1999, Ch. 1).

both at the same time), how are we to enable users to search for only the documents containing this meaning of *bass*?
- The term *guitar* only appears once in Document 1 — but it's very possible that a user searching for the term *guitar* would also find articles containing the term *Les Paul* relevant, since *Les Paul* is a make of *guitar*.

Many of these questions will be addressed during the rest of this book — though in truth, none of them can honestly be said to have been completely solved.

Now, here's the point. The rows of numbers in Table 5.1 can be thought of as vectors, just like the lists of numbers in the 'currency vectors' of Section 5.1. The different rows can be added together component by component, or scaled by any other real number, and it is a simple matter to check that these rows, and the operations of row addition and scalar multiplication, satisfy the axioms of Definition 5.5. Because of this, some of the theory and techniques of vectors, which are very well-developed and well-understood parts of mathematics, can be used to calculate and reason with words.

In fact, because they only have 3 coordinates each, it's almost possible to imagine these particular word vectors as points in a 3 dimensional space, as I've tried to depict in Figure 5.9. However, as we go through this section you'll realize that this diagram is only a visual aid — all of the mathematics we use to work out which words are similar to which other words can be done entirely by working with the coordinates in Table 5.1.

5.5 Similarity and distance functions for vectors

In this section we describe the most prominent ways to measure similarity and distance between vectors with any number of dimensions, and show how they can be applied to our word vectors. This should provide you with some robust mathematical tools and terminology (which may seem challenging, but which are easy to use in practice and very easy to program into a computer) and a taste of the possible linguistic applications.

We start with a few examples from the term-document matrix in Table 5.1, which gives word vectors such as

$$\textit{bank} = (0, 0, 4), \quad \textit{bass} = (2, 4, 0) \quad \text{and} \quad \textit{money} = (0, 1, 2).$$

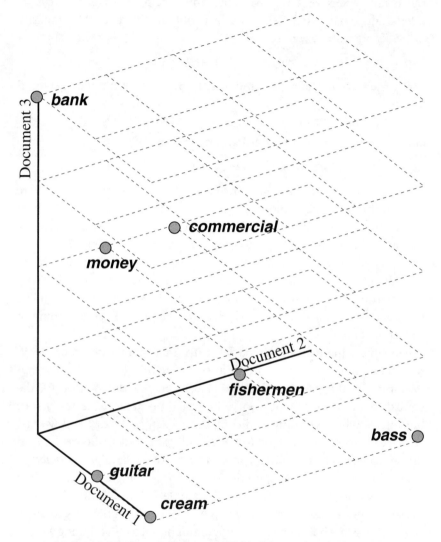

FIGURE 5.9 Imagining the word vectors of Table 5.1 as points in the three-dimensional space $\mathbb{R}^3$

In order to measure similarity between these word vectors, we could just multiply their first, second and third coordinates together and add the results, exactly as we did when introducing the cosine measure (Definition 7, page 106). For example, to obtain a score for the similarity of *bank* and *money*, we calculate

$$\text{sim}(\textit{bank}, \textit{money}) = (0 \times 0) + (0 \times 1) + (4 \times 2) = 8.$$

Using the same measure, the similarity between *bass* and *money* would be

$$\text{sim}(\textit{bass}, \textit{money}) = (2 \times 0) + (4 \times 1) + (0 \times 2) = 4,$$

and the similarity between *bank* and *bass* would be

$$\text{sim}(\textit{bank}, \textit{bass}) = (0 \times 2) + (0 \times 4) + (4 \times 0) = 0.$$

If this doesn't make sense straight away, do pause to work out where these numbers have come from and why they've been paired up in this fashion. The resulting 'similarity scores' are not without merit — for these three words, they show us that the two most similar words are *bank* and *money*, that *bass* and *money* are less similar though not unrelated (which in this context appears to make sense — the similarity is drawn entirely from Document 2, which does discuss the financial interests behind bass fishing), whereas *bass* and *bank* have nothing in common at all.

However, there are drawbacks with this similarity score — in particular, it gives higher similarities to more frequent words. For example, for the vectors *guitar* $= (1, 0, 0)$ and *cream* $= (2, 0, 0)$, the similarities for *cream* will always be exactly double those for *guitar*: but just as when we compared composers such as *mozart* and *wagner* with composers such as *hottenterre* and *giamberti* in Section 4.3, it doesn't seem right that *cream* should be twice as 'similar' as *guitar* to every single word, just because it occurs twice as often. (Of course, the distribution in our three document example is hardly realistic, but it makes the general point that frequent words might get higher similarity scores across the board, and this would not be such a good thing.)

Just as in Section 4.3, we can get round this problem by *normalizing* or dividing out our similarity scores. This will be easiest if we introduce some notation and terminology for vectors in general, which will also be frequently used in later chapters.

5.5.1 Introducing the vector space $\mathbb{R}^n$

So far we've seen that a journey vector a in the plane may be given by two coordinates. Because each point is given by 2 real numbers, the plane is given the name $\mathbb{R}^2$. Because our word vectors in Table 5.1 are collected from 3 different documents, each word vector in this example has 3 coordinates. The space of all possible collections of 3 real numbers is given the name $\mathbb{R}^3$. This idea can easily be generalized — a vector with n real number coordinates is considered to be a point in $\mathbb{R}^n$.

Definition 12 The vector space $\mathbb{R}^n$ is made up of all lists of the form $(a_1, a_2, \ldots, a_n)$ where each of these entries is a real number.

By definition, each point in $\mathbb{R}^n$ has n coordinates, so the dimension of $\mathbb{R}^n$ is the number n.

> **Arrays**
>
> The concepts and the notation behind the vector space $\mathbb{R}^n$ will probably be familiar to any programmer, even though the terminology may be different. A vector is just the sort of data that can be stored as an *array*, and a vector in $\mathbb{R}^n$ is an array whose elements are n 'real' numbers (in practice, these real numbers are approximations such as the floating point numbers and doubles of the C language).
>
> If arr is an array variable then its jth element is usually referred to by arr[j], which is very similar to the subscript notation of Definition 12, where a_j is the jth coordinate of the vector a. The only difference is that in computing, the first element of an array is usually called arr[0] rather than arr[1].

Addition of vectors in $\mathbb{R}^n$ is defined just as you would expect — the sum of two vectors $(a_1, a_2, \ldots, a_n)$ and $(b_1, b_2, \ldots, b_n)$ is

$$(a_1 + b_1, a_2 + b_2, \ldots, a_n + b_n).$$

To multiply the vector $(a_1, a_2, \ldots, a_n)$ by a given number or 'scale factor' λ, again multiply each of the coordinates in turn, giving

$$(\lambda a_1, \lambda a_2, \ldots, \lambda a_n).$$

This may look horrible. If you're having difficulty, go back to the calculations with currency vectors in Section 5.1, which are precisely examples of the equations above. Because we're so used to the idea of currencies, it's completely natural to think of adding two collections together by adding the dollars to the dollars, then adding the pounds to the pounds, etc. It also makes sense that to scale a whole currency vector by a given scale factor, you have to scale each currency (or each coordinate) by this factor in turn. The equations we've just written down are nothing more than general versions of exactly these 'one currency at a time' operations.

Once we start dealing with more than two or three coordinates, we normally stop using different letters such as (x, y, z) for the different coordinates, for two reasons. The first is that sooner or later we'd run out of letters. The second is that it's actually *easier* to use the subscript notation of Definition 12. How it will normally work is that we'll use one letter for each whole vector (such as a), and subscripts such as a_1, a_2, etc. for the coordinates. In this way, there are fewer letters to keep track of, and you know straight away that (for example) a_3 means "the 3rd coordinate of the vector a." This sort of notation makes it very easy to extend the Euclidean distance and cosine similarity measure of Chapter 4 to cope with any number of coordinates.

5.5.2 The scalar product of two vectors

Now that we're becoming familiar with vectors in $\mathbb{R}^n$, and the way of writing the coordinates $(a_1, a_2, \ldots, a_n)$ for the vector a, we can use these techniques to derive equations for computing similarities and distances for general vectors. We've already used a form of similarity score between our word vectors, by multiplying their coordinates and adding the results. This 'product' of two vectors has a special name.

Definition 13 Let $a = (a_1, a_2, \ldots, a_n)$ and $b = (b_1, b_2, \ldots, b_n)$ be vectors in $\mathbb{R}^n$. Their *scalar product* $a \cdot b$ is given by the formula

$$a \cdot b = a_1 b_1 + a_2 b_2 + \ldots + a_n b_n.$$

The scalar product is sometimes called the *inner product* or just the *dot product* of two vectors. We'll see shortly that the scalar product is closely linked to both the Euclidean distance and cosine similarity measures of Chapter 4.

5.5.3 Euclidean distance on $\mathbb{R}^n$ — extending Pythagoras' theorem to n dimensions

The two-dimensional version of Pythagoras' theorem (Section 101) will already be familiar to most readers because it is widely taught in schools. This enables you to work out the length of the hypotenuse of a right-angled triangle (alternatively, the length of the diagonal of a rectangle).

A less well-known but equally interesting fact is that exactly the same principle can be used in *three* dimensions to calculate the length of the diagonal of a solid box or 'cuboid' (which is the three dimensional version of a rectangle). For example, if the three perpendicular sides of a box have lengths p, q and r, then the diagonal of the box will have a length of $\sqrt{p^2 + q^2 + r^2}$, which is easy to prove.[7] If the corners of the box are the points a and b, which have coordinates (a_1, a_2, a_3) and (b_1, b_2, b_3) respectively, then the lengths of these sides are $p = b_1 - a_1$, $q = b_2 - a_2$ and $r = b_3 - a_3$. The length of the diagonal, which is the distance $d(a, b)$ between a and b, is therefore given by the formula

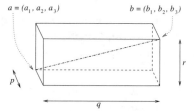

FIGURE 5.10 Measuring the diagonal from a to b.

$$d(a, b) = \sqrt{(b_1 - a_1)^2 + (b_2 - a_2)^2 + (b_3 - a_3)^2}. \qquad (5.10)$$

This is essentially the same as the two-dimensional distance formula given in equation (4.6), though we've changed some of the notation. We can also use the summation sign Σ (the capital Greek letter sigma) to mean "the sum of all these numbers," in which case equation (5.10) can be written using the shorthand

$$d(a, b) = \sqrt{\left(\sum (b_i - a_i)^2\right)}, \qquad (5.11)$$

where the subscript i stands for each of the indices in turn. This means exactly the same thing as equation (5.10), once you've expanded out the terms in the $\sum$ expression. (If we wanted to make it particularly explicit

---

[7]The proof works by applying Pythagoras' theorem twice: first consider the diagonal of the rectangular base of the cuboid, whose length is $\sqrt{p^2 + q^2}$; and then consider the hypotenuse of the triangle made by this diagonal and the upright distance r.

that the subscript i takes the values 1, 2 and 3 we'd write $\sum_{i=1}^{3}$ though normally this will be unnecessary because it should be clear from the context what range of values the subscript i takes.)

The same index and summation notation we used in equation (5.11) can be used to express the scalar product of two arbitrary vectors $a = (a_1, a_2, \ldots, a_n)$ and $b = (b_1, b_2, \ldots, b_n)$ as

$$a \cdot b = \sum a_i b_i.$$

This expression is much more succinct than the first version in Definition 13, and in the long run this brevity of expression is a great benefit. (It's a bit like using acronyms such as USA and UN for special terms — it takes a newcomer a while to get used to, but once you've got it, the idea of writing out the full expression in words every time is far too cumbersome for real life.)

Example 7 Calculating Distances

To make this a bit easier to grasp, we'll use equation 5.11 to work out the distances between some of our word vectors in Table 5.1, which contains the word vectors

$$\textit{bank} = (0, 0, 4), \quad \textit{guitar} = (1, 0, 0) \quad \text{and} \quad \textit{money} = (0, 1, 2).$$

Applying equation 5.11 gives

$$\begin{aligned} d(\textit{bank}, \textit{money}) &= \sqrt{(0-0)^2 + (0-1)^2 + (4-2)^2} \\ &= \sqrt{0 + 1 + 4} \\ &= \sqrt{5} \\ &\approx 2.24. \end{aligned}$$

In the same way, we get

$$d(\textit{bank}, \textit{guitar}) = \sqrt{1 + 16} = \sqrt{17} \approx 4.12.$$

Comparing these two results, we find that *guitar* is 'further' from *bank* than *money* is — a reasonable enough deduction.

However, there are problems with this technique — large vectors tend to be more distant from most other vectors than small vectors. For example, yet more calculations using equation 5.11 tell us that

$$d(\textit{guitar}, \textit{bass}) \approx 4.12 \quad \text{whereas} \quad d(\textit{guitar}, \textit{fishermen}) \approx 3.16.$$

You can confirm this by looking again at Figure 5.9, where *guitar* is slightly closer to *fishermen* than to *bass*. But this isn't really because *guitar* and *fishermen* have more in common than *guitar* and *bass* — a brief glance at Table 5.1 shows that *guitar* and *bass* have some coordinates in common whereas *guitar* and *fishermen* have none. Instead, the smaller distance between *guitar* and *fishermen* is simply because they're both nearer to the origin point. You can almost think of this as word vectors getting thrown out from a 'lexical supernova.' Each time a word is mentioned in the document collection it gets pushed futher from the origin point. The frequently occuring words get thrown the furthest and end up isolated in deep space, while the infrequent words don't get thrown very far and end up clustered around the origin, closer to one another.

Using subscripts for coordinates and the summation sign Σ takes a bit of mental gymnastics but it's well worthwhile because you can write down much more general formulas and equations without any extra effort. In particular, equation (5.11) can be used to define a distance function for a vector space of *any* dimension, without being changed at all. So far in this section, we've presumed that we're working in $\mathbb{R}^3$, where each vector a has 3 coordinates (a_1, a_2, a_3). But equation (5.11) works in just the same way for longer vectors $(a_1, a_2, \ldots, a_n)$, where the number of coordinates n can be as big as you like. In this way, Pythagoras' theorem can be adapted to give a distance measure on the vector space $\mathbb{R}^n$ for any dimension n. This measure is called the *Euclidean distance* measure on $\mathbb{R}^n$, and since it obeys the three metric space axioms (Definition 6), it is sometimes called the *Euclidean metric*.

This is another example of the fairly relaxed mentality you need to accept the ways vectors and dimensions are used. The analogy by which Pythagoras' theorem is extended from 2 dimensions to 3 dimensions should be pretty clear: the next step of applying the same numerical formulas to more than 3 dimensions isn't complicated. However, the conceptual step of trying to imagine the Euclidean distance function in $\mathbb{R}^4$ as somehow measuring the length of the diagonal of a 4 dimensional box is challenging, to put it mildly, if not downright crazy. But you don't need to do this to understand word vectors, just as we don't really need the diagram in Figure 5.9 in order to understand the significance of the word vectors Table 5.1. The main point behind Descartes' contribution to this sort of mathematics is that *we don't need diagrams for everything* — we can work out the distances algebraically, straight from the numerical coordinates. This is the reason why these techniques can be so easily adapted to spaces with more than three dimensions which we can't visualize so easily.

5.5.4 Norms and unit vectors

We now know how to calculate the Euclidean distance and the scalar product between two vectors a and b. However, we've also seen that neither of these measures is an ideal way to work out similarities and distances between word vectors: with the Euclidean distance, frequently occurring words with large word vectors end up *too far* from most other words, and with the scalar product, the same frequently occurring words end up *too similar* to most other words. What we need is a way of factoring out these unfair advantages and disadvantages, just as we wanted to even out our graph similarity scores in Section 4.3.1.

To do this, we measure the size or length of each vector,

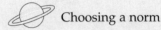 Choosing a norm

Choosing the right norm function for a given vector space will depend on your scientific purpose in building the space. For currency vectors, the only sensible way to find the value of a whole currency vector would be to use exchange rates. At the time of writing, £1 = $1.66, €1 = $1.18, and ¥1 = $0.0092, and so the vector $Dm = (6, 20, 15, 100)$ has the total value or norm

$$6 + 20 \times 1.66 + 15 \times 1.18 + 100 \times 0.0092 = \$57.82,$$

measured in dollars. In effect, we are using a 'weighted Manhattan metric' (page 102) to evaluate the length of a currency vector. (In practice, the norm of a currency vector should allow negative as well as positive values, in case you have more debts than assets — not all financial distances are positive.)

which is its distance from the zero point or origin. Applying equation (5.11) to the vectors $0 = (0, 0, \ldots, 0)$ and $a = (a_1, a_2, \ldots, a_n)$, we have

$$d(0, a) = \sqrt{\sum (a_i)^2}.$$

This can also be expressed in terms of the scalar product, since $\sum (a_i)^2$ is the same as $\sum a_i a_i$ which is just $a \cdot a$. This is summed up in the following definition.

WORD-VECTORS AND SEARCH ENGINES / 157

Definition 14 The *norm* or *length* of the vector a is written $||a||$ and is defined to be
$$||a|| = \sqrt{\sum(a_i^2)} = \sqrt{a \cdot a}.$$

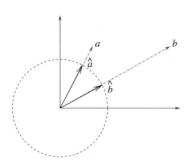

FIGURE 5.11 Normalized or 'unit' vectors

Just as the norm $||a||$ can be defined as the Euclidean distance $d(0, a)$, so also the Euclidean distance between two vectors a and b can be obtained by the formula $d(a, b) = ||b - a||$, which is the norm of the journey vector from a to b. So in many ways, whether we choose to talk in terms of norms or distances is a matter of convenience, just as whether we think of a vector as a point or as a journey from the origin to that point is a matter of convenience.

This norm is exactly the factor we need to divide by in order to remove any extra preference or penalty given to the word vectors of frequently occurring words. It's easy to check that if you divide every coordinate in a by the norm $||a||$, you are left with a vector whose norm or length is equal to 1. This vector is written as $\frac{a}{||a||}$ or sometimes $\hat{a}$, and is called the *unit vector* of a. In other words, $\hat{a}$ is the vector in the same direction as a whose length is just one unit. One simple way to think of the unit vector $\hat{a}$ is as the projection of the vector a onto a sphere or circle of radius 1, since all the points on this sphere or circle have a distance of one unit from the origin (Figure 5.11). Since all unit vectors have the same length, the only thing that distinguishes these vectors is their directions.

We can now replace all the word vectors in Table 5.1 with their unit vectors, giving the normalized version in Table 5.2. What we've done is found the norm of each row and divided each entry in that row by this number. To check that these normalized vectors *are* unit vectors, simply go along each row in the table, square each individual number and add together the results — you should get the answer 1 (or at least, as nearly as our approximation to 3 decimal places will allow).

TABLE 5.2 The vectors from Table 5.1 after they have been normalized

	Document 1	Document 2	Document 3
bank	0	0	1
bass	0.447	0.894	0
commercial	0	0.707	0.707
cream	1	0	0
guitar	1	0	0
fishermen	0	1	0
money	0	0.447	0.894

5.5.5 Cosine similarity in $\mathbb{R}^n$

The scalar product of two *normalized* vectors corresponds exactly with their *cosine similarity*, as defined in Section 4.2.1. We are thus in a very useful practical situation: we can work out similarities between words simply by working out the cosine similarities between their vectors in Table 5.2 — by multiplying together the corresponding coordinates and adding the results. For example, we now have

$$\cos(\textit{guitar}, \textit{cream}) = 1 + 0 + 0 = 1,$$
$$\cos(\textit{guitar}, \textit{bass}) = 0.477 + 0 + 0 = 0.477,$$

and

$$\cos(\textit{guitar}, \textit{fishermen}) = 0 + 0 + 0 = 0.$$

Now we have that *guitar* and *cream* are the most similar pair (which with the meaning of *cream* in Document 1 is what we should have), that *guitar* and *bass* have a certain amount in common (in fact, they have one of the meanings of *bass* in common but not both), and *guitar* and *fishermen* are completely unrelated. The skewing of such results because of *bass* being a longer vector, and recurrent problems of this nature, are gone for good.

The cosine similarity of any pair of vectors (not just unit vectors) can easily be obtained by taking their scalar product first and then dividing by their norms (rather than normalizing all vectors first and then computing cosine similarities between these normalized vectors). This is probably the most usual way of defining cosine similarity, and you will often see equations like

$$\cos(a, b) = \frac{a \cdot b}{||a|| \, ||b||} \qquad (5.12)$$

used in the literature to define the cosine similarity between two vectors a and b. As we said in Section 4.2.1, this number can also be thought of as the cosine of the angle between the vectors a and b, which is a good way to think of this measure of similarity which makes it intuitively clear that we are only interested in the directions, not the lengths of the vectors (since the angle between two lines doesn't depend in any way on the lengths of those lines).

To see that this way of measuring the similarity between normalized vectors is a reasonable approximation to measuring the similarity between words, have a good look at Table 5.3. This finally gives the similarity between *each* pair of word vectors. Remember that this is all calculated from the fragment of language contained in Documents 1, 2 and 3 in Figure 5.8, and in this tiny sample many of the words only occur with a fraction of their possible meanings. Given this, Table 5.3 does a remarkably good job of modelling which words are similar and dissimilar to one another.

Such a table of similarities between each pair of objects in a collection can be called a *data matrix* or an *adjacency matrix*. Note that the main top-left to bottom-right diagonal is made entirely of 1's — this is because each word has a similarity of 1 with itself, because the length of each normalized vector is equal to 1. If we reflect the table about this diagonal (thereby interchanging the rows and the columns), notice that we get the *same* table. A table (or matrix) with this property is called *symmetric*,

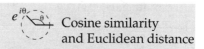

Cosine similarity and Euclidean distance

The cosine similarity and Euclidean distance between two unit vectors are closely related: if a and b are unit vectors then it follows that

$$\begin{aligned}(d(a,b))^2 &= ||b-a||^2 \\ &= (b-a)\cdot(b-a) \\ &= b\cdot b + a\cdot a - 2a\cdot b \\ &= 2 - 2a\cdot b.\end{aligned}$$

It follows that the relative ranking of points as being 'more or less similar' according to cosine similarity or 'closer or further away' according to Euclidean distance will be the same for unit vectors.

However, because the left hand side in this equation is squared, cosine similarity is 'less transitive' than Euclidean distance. For example, the vectors $a = (1,0)$, $b = (\frac{\sqrt{2}}{2}, \frac{\sqrt{2}}{2})$ and $c = (0,1)$ all have length 1, and although $\cos(a,b) = \cos(b,c) = \frac{\sqrt{2}}{2} \approx 0.707$ which is quite high, $\cos(a,c)$ is still 0.

TABLE 5.3 The cosine similarities between each pair of words in our model

	bank	bass	commercial	cream	guitar	fishermen	money
bank	1	0	0.707	0	0	0	0.894
bass	0	1	0.632	0.447	0.447	0.894	0.400
commercial	0.707	0.632	1	0	0	0.707	0.915
cream	0	0.447	0	1	1	0	0
guitar	0	0.447	0	1	1	0	0
fishermen	0	0.894	0.707	0	0	1	0.447
money	0.894	0.400	0.915	0	0	0.447	1

because it is symmetrical in the traditional "it looks the same if you reflect it in a mirror" fashion, and because of this geometric symmetry, these matrices can be used to represent symmetric *relations* (page 60).

5.6 Document vectors and information retrieval

This chapter was meant to tell you how a search engine works, and so far it's been entirely about *words* and these things called vectors, whereas one thing that all readers will probably know is that search engines are about finding *documents*. In case you're feeling tricked into reading a whole chapter on mathematics under false pretences, we will finish this chapter by describing how to create *document vectors* so that we finally have a little system which can take a query made out of any combination of the seven terms in Table 5.3 and rank the three documents in Figure 5.8 according to their relevance to such a query.

Norms and Euclidean distances of vectors are just as easy to adapt to n dimensions: in some ways, computers find this easier than people because computers don't care whether or not they can visualize the vectors they're using.

The trick is very simple: we represent the documents as vectors in *the same space* as the words, and then we can compute similarities between words and documents just as we computed similarities between pairs of words. In a sense, we've already been doing this — our word vectors have had three coordinates, one for their occurrence in each of the documents, and in this sense the three documents have been used as a *basis* for our space of word vectors. One way to make this clearer is to reason that if we had 4 documents in our collection, we would need 4 columns in the term-document matrix and each word vector would have 4 coordinates, and so on for n documents: so the number of documents clearly determines the dimension of the space of word vectors. Another way to see this pictorially is to consult the diagram in Figure 5.9, where the 3 documents are clearly being used as 3 coordinate axes. Representing each document as a unit vector in the direction of its coordinate axis, we have the *document vectors*

$$Doc1 = (1, 0, 0), \quad Doc2 = (0, 1, 0) \quad \text{and} \quad Doc3 = (0, 0, 1).$$

It's now a simple matter to work out the cosine similarity between a given query expression and each document in turn. For example, for the query *bass* with vector $(0.477, 0.894, 0)$, we get

$$\cos(\textit{bass}, \textit{Doc1}) = 0.477, \qquad \cos(\textit{bass}, \textit{Doc2}) = 0.894,$$
$$\text{and} \qquad \cos(\textit{bass}, \textit{Doc3}) = 0.$$

If the user was returned the winning document *Doc2* and realized that this isn't the sort of *bass* they were looking for, they could add the term *guitar* to give the query *bass guitar*. Now the benefits of using vectors really begin to pay off — we can simply add together the vectors for *bass* and *guitar* to give a new *query vector*,

$$\textit{bass} + \textit{guitar} = (0.477, 0.894, 0) + (1, 0, 0) = (1.477, 0.894, 0),$$

which when normalized becomes $(0.855, 0.518, 0)$.

Comparing each document with this new query vector, *Doc1* is the winner with a cosine score of 0.855. In a very simple way, we have combined the meanings of two words to give a meaning for a combination of words, for the first time in this book. This very important process is called *semantic composition*, which we are modelling here (very bluntly) using vector addition.

You can carry on playing the game of working out query vectors, comparing them with the three document vectors, and see if the ranking you get coincides with how relevant the documents are to your query.

That is one basic way to build an information retrieval system, though there are many important questions to ponder. Many researchers and engineers have improved over this baseline, finding out how to assess the importance of different terms and documents (Google's *PageRank* is one example of an answer to such a question), how the vector model compares with other conceptual models, how different terms might depend on one another, and how to cope with the engineering task of managing bigger and bigger document collections and term-document matrices. The wider reading section at the end of this chapter will barely be able to scratch the surface of these topics, though it will try to give some good leads which you can follow up for yourself.

This may seem like a lot of mathematics in order to end up multiplying a few numbers here and there and coming up with a ranking of our three documents. In many ways, this is one of the vector model's great strengths — the mathematics in this chapter may have seemed to have some strange names and symbols and far too many dimensions, but at the end of the day this is nothing more than a shorthand for adding and multiplying lots of different numbers. These numbers can be extracted straight from a document collection without any

> **Programming cosine similarity in higher dimensions**
>
> If we use arrays to hold our vectors and a `for` loop to work out the cosine, cosine similarity in n dimensions can be implemented using exactly the same code as for 2 dimensions. If your (normalized) vectors are called `vec1` and `vec2` and their cosine similarity is stored by the variable `cos`, the code would read something like
>
> ```
> cos=0;
> for( i=0; i < dim; i++){
>   cos=cos+vec1[i]*vec2[i];
> }
> ```

need for deep linguistic analysis, which is one of the main reasons that information retrieval has proved to be such a widespread and successful branch of natural language processing: your system really doesn't need to know much about *language* at all. (If you just stop for a moment to think about how much more trouble we'd have had in building a simple system that could read out a spoken version of three documents or translate them into a different language getting both the grammar and the meaning correct, I think you'll agree that building an information retrieval system was pretty easy.)

Another vital point is that the toy system we've described in this chapter is *scalable*. This is where the mental gymnastics of working in many dimensions pays off. It may have seemed like unnecessarily hard work to develop all this n dimensional mathematics when most of the time we were just working in 2, 3 and 4 dimensions. But towards the end of the chapter we realized that we were using a dimension for each separate document. Now, if we'd confined ourselves to a comfortable mathematical system that could just cope with 2 or 3 dimensions, we'd have no hope of coping with a bigger document collection. But instead, we put in the hard work to use a mathematical language that can be adapted to *any* number of dimensions: so the same system can be used for a document collection of any size.[8]

I'd like to finish this chapter with two observations which take us back to the concepts of Chapter 1. We said in Section 1.5.4 that the 'measurements' we made of the extent to which different words occurred in different relationships or contexts would be *discrete*, but that many of the models we would build from this information would rely on *continuous* mathematical techniques. The vector spaces of this chapter are precisely the sort of continuous model to which we were referring. We started by counting discrete whole numbers (for example, in Table 5.1), but by the time we had normalized these vectors in Table 5.2, we needed all sorts of numbers in between 0 and 1, and in principle the only limit on the specificity of these numbers is how many decimal places it's worthwhile keeping for each of them. When we're just comparing the relevance of three different documents, it doesn't matter so much: when we're dealing with a document collection the size of the World Wide Web, we may need more and more in-between numbers to keep track of ever finer distinctions of importance and relevance. The binary partition into relevant and non-relevant documents could not support a user who needed to know which out of thousands of relevant documents were the *most* relevant, and this challenge has naturally led researchers to

---

[8]In the presentation in this chapter, I have departed slightly from the approach in most textbooks, which uses the words or *terms* as coordinate axes from which to represent both document and query vectors. Instead I have used the documents as coordinate axes for repreenting the terms, because this forms a better introduction to the way we will use word vectors in the rest of the book, where we are much more interested in computing word-word similarities than (say) document-document similarities. Also, it was much easier to try to draw Figure 5.9 as 7 word points in 3 document dimensions than as 3 document points in 7 word dimensions! Some mathematical issues related to such questions are explored by Yang et al. (1998) in the context of translingual information retrieval.

study continuous methods such as vector spaces, fuzzy set theory and probability.

Continuous methods enable us to model not only atoms of meaning such as words, but the space or void in between these words. Whole passages of text are mapped to points of their own in this void, without changing the underlying shape of the space around them, and we can measure the distance between query terms and these document points very easily, if naïvely.

Our models have come all the way from Chapter 2, where we studied relationships between words almost without using any numbers at all, to the work in this chapter where words have been represented using *only* numbers, obtained by measuring how many times a word occurs in different documents. Geometric comparisons such as longer or shorter, closer or further away, have been made purely by doing calculations with these coordinates. This has all been made possible by the 'Cartesian revolution' in mathematics which enabled geometric problems to be described and solved using relationships between collections of numbers.

Wider Reading

The introduction to vectors given here is necessarily brief and very informal. Those who wish to go deeper or more thoroughly into any of the concepts introduced here should consult a proper book on linear algebra. Chapters 2 and 3 of Jänich (1994) are appropriate and readable: another good introduction is Vallejo (1993).

The rest of this book will describe many important operations with vectors, and the ways they have been used for analyzing linguistic information: a lengthy list of wider reading on this topic will gradually emerge. Those interested in the particular uses of vectors for information retrieval should consult the classic text of Salton and McGill (1983, Ch 4), and the summary contained in Jurafsky and Martin (2000, §17.3) which gives a good pictorial overview.

The principal mathematical models used for information retrieval systems are the *vector*, *probabilistic* and *fuzzy set* models, all of which are discussed in Baeza-Yates and Ribiero-Neto (1999). The fuzzy set model is an adaptation of the traditional Boolean model (which will be described in Section 7.1) to cope with continuous values. Fuzzy sets for information retrieval are discussed at length by Miyamoto (1990), starting with a good introduction to what fuzzy sets are, and providing fuzzy set models for some of the traditional thesaural relationships such as 'Broader Term' which we met in Chapter 2. One of the most interesting probabilistic models is the 'inference network' used in the INQUERY system (Turtle and Croft, 1989, Callan et al., 1992): such networks use probability theory to find important paths in directed graphs of the sort we studied in Chapter 3, in this case the path that leads between a query and relevant documents.

A unifying theme behind these three models is that normalized vectors, probability distributions and fuzzy sets are all mathematical techniques which replace binary values (0 or 1) which signify 'not belonging' or 'belonging' with continuous values (0 or 1 or anything in between) which signify 'partially belonging' or 'probably belonging.' There are some practical differences between the models (for probability distributions, the sum of all the individual probabilities must be equal to 1, whereas as we've seen for normalized vectors, the sum of the *squares* of the individual coordinates must be equal to one, at least with the Euclidean norm), but it seems at least possible that these are really different *implementations* of one underlying *model*.

Other important topics in information retrieval include weighting strategies, user models and interfaces, linguistic operations such as stemming words (such as *fishermen*) to their lexical roots (such as *fisherman* or even *fish*) and of course engineering. Rather than try to give piecemeal and unsatisfactory pointers to each of these topics, I recommend consulting the indexes of general books such as Salton and McGill (1983), Kowalski (1997) and Baeza-Yates and Ribiero-Neto (1999), and also the excellent range of papers collected by Sparck Jones and Willett (1997).

Those interested in the mathematical development of vectors should certainly look at the pioneering work of Descartes (1637). The use of numbers and lines together to solve problems is pioneered in Book I, which in particular introduces the idea of describing points by measuring their coordinates on two chosen lines (p. 29, p. 310 in the original French). There is a particularly beautiful version available from Dover which contains a facsimile of Descartes' original French manuscript next to the English translation. An excerpt from the vital first book of the Analytic Geometry is contained in Smith (1929, p. 397-403), and a good historical account can be found in Boyer and Merzbach (1991, Ch. 17).

The technique of using vectors to represent points in space grew from the work of Sir William Rowan Hamilton (1805-1865) on quaternions, in which he first describes the addition of a fourth dimension to a mathematical system (Hamilton, 1847). At the same time, during the 1840s, a little known German high school teacher called Hermann Grassmann (1809-1877) was developing a theory of *Ausdehnungslehre* (*extension theory*), and this work contains our modern notion of vectors in *any* number of dimensions (Grassmann, 1862). Since his purpose was to represent a general concept with any number of dimensions, our word vectors owe much more to this foundation — and as we shall see in Chapter 7, Grassmann contributed some of the tools which are key to adapting the power of logic to the geometric setting of a vector space.

In cognitive science, the use of coordinates and dimensions to model mental processes is the cornerstone of the *Conceptual Spaces* of Gärdenfors (2000). In particular, the first chapter of this book contains an excellent introduction to the notion of dimensions, and the way different stimuli such as taste and colour can be represented as points in such a conceptual space. For example, colours seem to have a persistent 3 dimensionality about them, whether those dimensions are 'red, green and blue' or 'hue, brightness and chromaticity.'

No bibliography about dimensions would be complete without mentioning Edwin Abbott's novel *Flatland*, originally published in 1884. This slim book (under 100 pages) is the darling of every enthusiast who's ever read it and the Dover edition will probably cost less than your bus fare to the bookshop to buy it. In Abbott's story (also a barbed satire against the hierarchical rigidity of Victorian society) the world of a humble square is rocked by the intersection of a solid sphere with his two-dimensional existence. At first resistant, the square becomes convinced that a *third dimension* is possible, and through trying to explain this to the other polygons he becomes a prophet, a heretic and finally a prisoner, clinging forlornly to the hope that his teachings will one day inspire a new generation "who shall refuse to be confined to limited Dimensionality."

6

Exploring Vector Spaces in One and More Languages

This chapter is *much* easier than the last one. Having gone through the mathematics and (in a small example) the process of building word vectors, this chapter concentrates entirely on the reward for these achievements: being able to learn a lot about words themselves by exploring their vector spaces. The tour guide nature of this book is being given free-rein: you've put in (or adroitly skipped!) the hard mathematical labour to get to this point, and now you get to be a tourist and enjoy the view.

Not that this chapter is without new ideas. We'll talk about different ways of making the views more informative than a simple list of related words, by grouping words into clusters or plotting their vectors in a two-dimensional 'word-spectrum.' We'll also see how to use documents from two languages to build a single vector space with words from *both* languages, which can (for example) be used to translate words and queries between languages.

While this hopefully makes a cohesive and readable story, you don't have to read it in sequence, and a perfectly acceptable way to approach this chapter is to flick through, see which of the pictures interest you, and delve into those sections to see how they were made. Or maybe even better, follow the links that I've highlighted at the beginning of sections and explore the models online for yourself. They say that "seeing is believing," and while the examples chosen for this chapter work especially well at highlighting particular points, there's no better way to convince yourself that the techniques we've described really do

TABLE 6.1 The nearest words to *fire* in models built from the BNC and the WSJ

fire, BNC		fire, WSJ	
fire	1.000000	fire	1.000000
firefighters	0.743202	killed	0.600383
blaze	0.673635	killing	0.575034
cease	0.647473	sarajevo	0.573097
fires	0.619991	serb	0.564908
flames	0.571786	civilians	0.552733
gunfire	0.567285	artillery	0.534770
burned	0.541329	serbs	0.528852
rescuers	0.525994	grozny	0.520625
wounded	0.514218	guard	0.516758
firefighter	0.509511	hostages	0.511710
evacuation	0.496393	damaged	0.509093
sniper	0.493552	injured	0.504811
mortar	0.490172	attacked	0.504070
lt	0.489406	injuring	0.503863
brush	0.485589	troops	0.503337

work in general than to use them for yourself. A few examples of your own will also give you a much better feel for what's going on, and with these under your belt, the mathematical methods described will make a lot more intuitive sense.

6.1 Welcome to WORDSPACE

Demo: `http://infomap.stanford.edu/webdemo`

Let's begin just by seeing the nearest neighbours of a few word vectors. You can do this just by typing in a word and clicking on "Send," as shown in Figure 6.1, which shows the neighbours of the word *fire* in vector models built from the British National Corpus (BNC) and the Wall Street Journal (WSJ). To make the results more readable, these neighbours are also shown in Table 6.1.[1]

You can see immediately that these words are pretty intelligent choices for a machine asked to give you words whose meanings are related to *fire*. You can also see that the models built from the two

---
[1] For the sake of clarity, many of the results will from now on be presented as Tables rather than screenshots. Nearly all the examples can be easily produced using the online demos, so please feel free to recreate them or at least imagine them with the comforting look and feel of a good web browser.

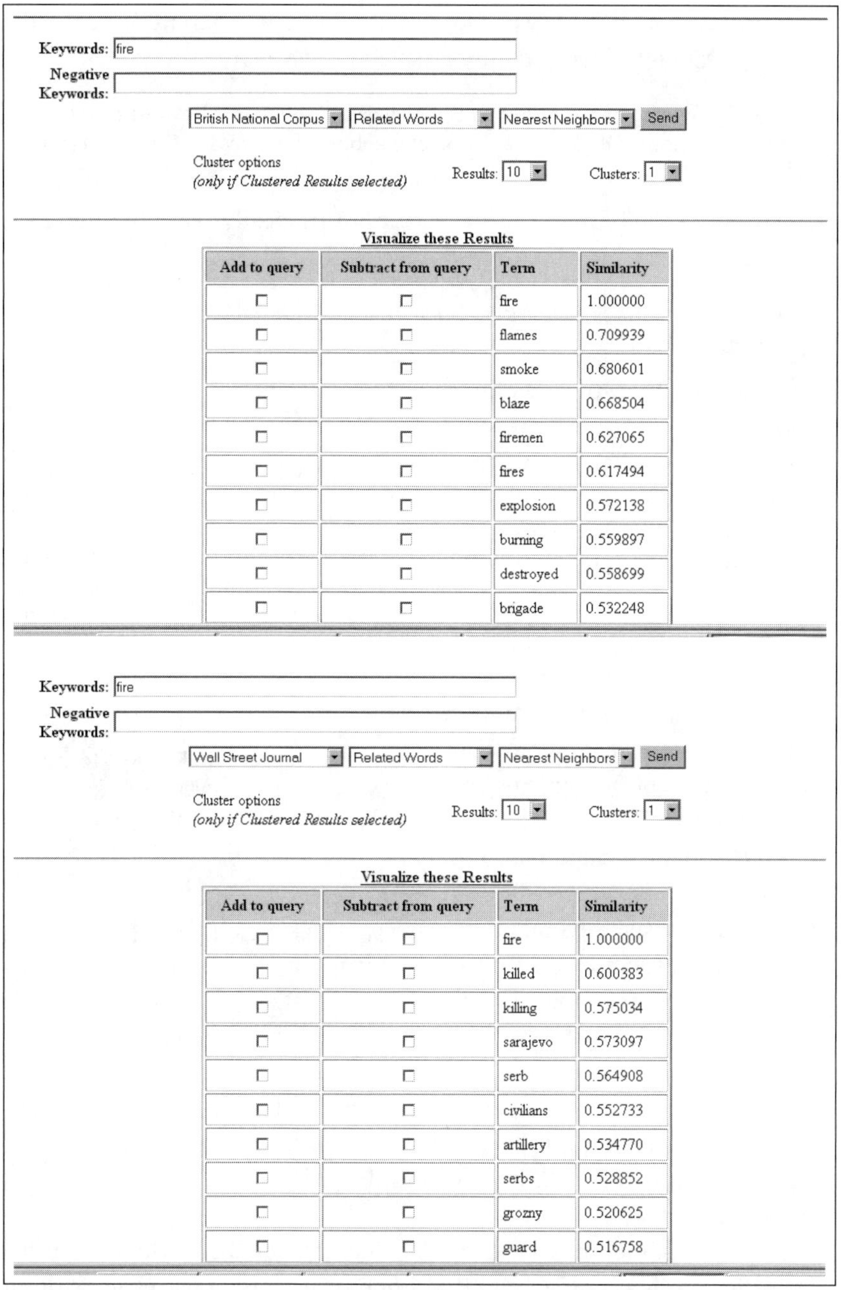

Keywords: fire
Negative Keywords:

British National Corpus ▼ Related Words ▼ Nearest Neighbors ▼ Send

Cluster options
(only if Clustered Results selected)   Results: 10 ▼   Clusters: 1 ▼

Visualize these Results

Add to query	Subtract from query	Term	Similarity
□	□	fire	1.000000
□	□	flames	0.709939
□	□	smoke	0.680601
□	□	blaze	0.668504
□	□	firemen	0.627065
□	□	fires	0.617494
□	□	explosion	0.572138
□	□	burning	0.559897
□	□	destroyed	0.558699
□	□	brigade	0.532248

Keywords: fire
Negative Keywords:

Wall Street Journal ▼ Related Words ▼ Nearest Neighbors ▼ Send

Cluster options
(only if Clustered Results selected)   Results: 10 ▼   Clusters: 1 ▼

Visualize these Results

Add to query	Subtract from query	Term	Similarity
□	□	fire	1.000000
□	□	killed	0.600383
□	□	killing	0.575034
□	□	sarajevo	0.573097
□	□	serb	0.564908
□	□	civilians	0.552733
□	□	artillery	0.534770
□	□	serbs	0.528852
□	□	grozny	0.520625
□	□	guard	0.516758

FIGURE 6.1 Screen shot of demonstration running at `infomap.stanford.edu/webdemo`

corpora use the word *fire* predominantly in very different contexts: the British National Corpus mainly talks about *fire* as in

> The **fire** brigade was called to Spencerbeck House in Middlesbrough in the early hours of yesterday morning where there was a **fire** in the bin room. (BNC)

whereas the Wall Street Journal has more tragically warlike references such as

> Twenty-six people, including 15 Americans, were killed in the "friendly **fire**" incident. (WSJ)

As well as telling us something about the use of the word *fire*, this immediately suggests another inference about the difference between these two corpora: it appears likely that the BNC contains more information on domestic topics, and the WSJ contains more information on international topics, and a brief glance at a few articles from both corpora confirms this hypothesis.

The online demonstration (shown in the screenshot in Figure 6.1) has several more options and functions, some of which will be described in this chapter. To begin with, we can

> **WORDSPACE software**
>
> The software used to build the vector spaces in this chapter is now freely available for researchers and commercial users, and can be downloaded from `http://infomap-nlp.sourceforge.net`
>
> The software enables you to build WORDSPACE models for words and documents from your own corpora, and has worked successfully in Linux, UNIX, Mac and Windows environments.
>
> Particular credit and personal thanks go to my colleague Scott Cederberg for making this release possible.

enter several words to give a compound query, whereupon the search engine adds their word vectors together just as in Section 5.6. For example, by changing the queries from *fire* to *enemy fire* in the BNC or *fire damage* in the WSJ, we can find neighbours related to the other senses of *fire* (Table 6.2).

This can sometimes be done by clicking on the "Add to Query" boxes in the demo, which will add one of the previous results so that it becomes part of the query, but there's a catch with this process: only

TABLE 6.2 Altering these queries to *enemy fire* and *fire damage* reverses the senses of *fire* compared with those in Table 6.1.

enemy fire, BNC		fire damage, WSJ	
fire	0.801632	fire	0.811863
enemy	0.801632	damage	0.811863
civilians	0.592632	storm	0.661878
wounded	0.583231	damaged	0.629903
casualties	0.577208	killed	0.588321
firefighters	0.573378	injured	0.588272
soldier	0.567039	quake	0.574127
battlefield	0.555124	severe	0.548609
cease	0.547627	accident	0.547943
marine	0.535930	causing	0.535091
soldiers	0.534775	devastated	0.518272
sniper	0.525046	fires	0.518011
deadly	0.522067	accidents	0.515751
forces	0.518783	contaminated	0.514275
confrontation	0.511486	hurricane	0.512288
army	0.508463	caused	0.511290

the top few results can be added in this way, and if a word you wanted is *already* among the top few results it's likely that you're already pretty happy with the area of meaning you're exploring. It can be much more effective to *remove* unwanted terms if the top few results are *not* what you wanted, which can help you to explore genuinely different senses of your query terms. We'll discuss the vector logic behind this negation procedure in Chapter 7: it is quite an exciting new development.

The score in the right hand column is none other than the cosine similarity between the query statement and this particular word vector. The panel at the top allows you to choose between different corpora and different search options: from the corpora you can retrieve documents as well as related words, and there are options to return clusters of related words instead of single words, and a link to a visualization panel enabling you to see a 2 dimensional 'spectrum' of results — there will be more about these functions later in the chapter.

6.2 Building a WORDSPACE

In this section we'll go into some more detail about precisely how these vectors were built from corpora. To begin with, we'll define a term we'll be using for the rest of this book.

Definition 15 A space of word vectors will be called a WORDSPACE.

The term WORDSPACE was introduced by Hinrich Schütze (1997), whose work at Stanford paved the way for some of our subsequent successes. Of course, the graphs and hierarchies we discussed in earlier chapters could also be described as 'word spaces,' but the term WORDSPACE will be used especially to refer to one of our vector spaces because this is common in the literature and because 'word vector space' is a bit of a mouthful.

There are at least two potential problems with a standard term-document matrix that can make it difficult to measure the similarities between words:

- In different corpora, documents are of very different lengths and consistencies.
- Most of the entries in a standard term-document matrix are usually zero, leading to very low cosine similarities between terms that might still have a lot in common.

We will consider these issues in turn, and how they are solved in Schütze's method for building a WORDSPACE.

6.2.1 Using cooccurrence with content-bearing words

Different corpora have very different ideas of what a document is. For the Ohsumed medical corpus (Hersh et al., 1994), each individual document is an abstract from an article in MEDLINE, so each document is a pretty short, coherent unit. However, for several Hansard[2] corpora (proceedings of different parliaments and government institutions), each separate document may simply contain a record of all issues that were discussed during a particular session, and many of these contain

---

[2]Luke Hansard (1752–1828), born in Norwich, was the founder of Hansard & Sons, a printing firm which specialized in printing the official reports and debates of the British Parliament. Hansard & Sons set such high standards that their work was replicated in many other places, and to this day, the records of parliamentary proceedings throughout the British Commonwealth are called *Hansards*.

TABLE 6.3 The cooccurrence of five different words with the content-bearing words *food* and *music*.

HOT-FROM-THE-OVEN MEALS: Keep **hot FOOD** **hot**; warm isn't good enough. Set the oven temperature at 140 degrees or hotter. Use a **meat** thermometer. And cover with foil to keep **FOOD** moist. **Eat** within two hours. (NYT, 1996)

"Change is always happening," said the ebullient trumpeter, whose words tumble out almost as fast as notes from his **trumpet**. "That's one of the wonderful things about **jazz MUSIC**." For many **jazz** fans, Ferguson is one of the wonderful things about **jazz MUSIC**. (NYT, 1996)

	music	food
eat	0	1
hot	0	2
jazz	3	0
meat	0	1
trumpet	1	0

a disparate variety of topics. In order to deduce (or at least, have a reasonable guess) that two words are related in some way, it's not enough just to know that they occur in the same document.

One good answer to this problem is to say that two words are likely to be related if they occur close together in text — for example, if they occur with fewer than 15 intervening words between them, or (if we have a way of detecting sentence boundaries) if they occur in the same sentence. Instead of building a table that records the occurrence of a word in a particular document, we build a table that records the cooccurrence of a pair of words near to one another.

If we were to do this for *every* pair of words, we'd end up with a huge, huge table. So instead, we select a comparatively small number of important *content-bearing words* which we hope are good indicators that a particular area of meaning is being talked about.[3] For example,

---

[3] In practice, we have done this by first discounting a 'stoplist' of very common words such as *if*, *of*, *the*, a standard process in information retrieval (Baeza-Yates and Ribiero-Neto, 1999, p. 167), since these words don't help very much when trying to capture what a piece of text is about. After these stopwords were removed we took the 51st to 1050th most frequent words as our 1000 content-bearing words.

Table 6.3 shows two excerpts from the New York Times, one of which is about *food* and the other about *music*. Suppose that *music* and *food* have been chosen as good content-bearing terms. Then the other words are assigned *food* and *music* coordinates depending on whether they occur with one of these two words.

Of course, mistakes can happen. If a content-bearing word occurs in a sentence, it doesn't necessarily mean that this sentence is *about* this particular content-bearing word, and we could easily be confused by a poetic sentence such as

> If music be the food of love, play on.

Often, frequent words can be some of the most ambiguous words in the language, so one might argue that using these as indicators of different semantic areas is very naïve. Nonetheless, after we've gone through several million words of text, finding which of the content-bearing words each of the other words cooccurs with, we end up with a very large table of numbers where, judging by the results, a lot of these idiosyncracies have evened themselves out. The final cooccurrence table will be an enormous version of the example in Table 6.3, containing one column (i.e. one dimension) for each content-bearing word and one row for each of the other words for which we want to build a word vector. In many of our experiments, we've used 1,000 columns

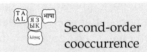
Second-order cooccurrence

Another benefit of using content-bearing words instead of documents as columns in our table is that many words which are roughly synonymous do not occur in the same documents. For example, both *referee* and *umpire* mean "the adjudicator of a game," but most games which have such a judge use either the term *referee* or the term *umpire* but not both. It follows that if one of these words is used in a report of a sporting event, the other does not usually appear.

If, instead of looking for occurrences of these words in the *same* documents, we measure the extent to which both these words cooccur with the same content-bearing words (such as *play* and *game*) in different parts of the corpus, we can still measure an accurate similarity between *referee* and *umpire* even if they *never* occur in the same document (Schütze, 1998).

(In spite of this, it's still very clear in the BNC WORDSPACE that *umpire* is much more similar to *cricket* and *referee* is much more similar to *football*.)

and 20,000 rows, giving us 20,000 word vectors in a 1,000-dimensional vector space.

6.2.2 Reducing Dimensions using Latent Semantic Analysis

A table with 1000 columns is still a pretty big table — there are a lot of zeros (or spaces) and also a lot of redundant dimensions. What do I mean by redundant dimensions? Well, suppose that two of the content-bearing terms in a model built from the BNC were *car* and *drive* (and if we've selected our content-bearing terms statistically by simply choosing frequent words, this is perfectly likely). Then many of the words that occur with *car* will also occur with *drive*, since sentences about cars and sentences about driving are normally about the same underlying topic. These underlying topics are the variables we really want to have as coordinate axes in our WORDSPACE, rather than the words which signify them. To model this intuition, we approximate this underlying variable by drawing a best-fit line through the data points, giving a new *latent axis*. Each word is then given a new coordinate along this latent axis by dropping a perpendicular from its original position onto the new coordinate axis, sometimes called *projecting* the word onto the new axis (Figure 6.2).

Of course, there are outliers that don't fit with the topic of cars and

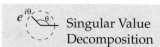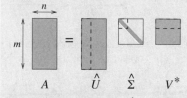

Singular Value Decomposition

The algebraic algorithm behind LSA is called *singular value decomposition* (Trefethen and Bau, 1997, Ch 4). Any $m \times n$ matrix A can be factorized as the product of three matrices $\hat{U}\hat{\Sigma}V^*$, where the columns of $\hat{U}$ are orthonormal, V^* is an orthonormal matrix, and $\hat{\Sigma}$ is a diagonal matrix.

The diagonal entries of $\hat{\Sigma}$ are called the *singular values*, and they can be arranged in nonincreasing order so that the biggest are at the top. If we only use the first k singular values and treat the rest as zero, the product $\hat{U}\hat{\Sigma}V^*$ uses only the left hand k columns of $\hat{U}$ and the top k rows of V^*. Each m-dimensional row vector A can thus be mapped to a k-dimensional 'reduced vector' in the k-dimensional subspace spanned by the top k rows of V^*.

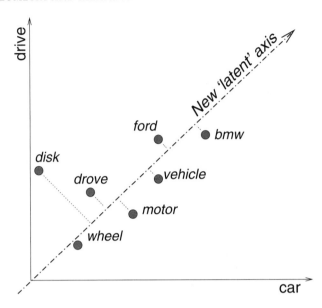

FIGURE 6.2 Projecting words related to cars and driving onto a new coordinate axis that combines both these variables.

driving — for example, a *disk drive* has nothing to do with cars at all. However, outliers such as this can actually be pretty well handled by the projection method. Firstly, note that the coordinate of the word *disk* in the new 'cars and driving' direction is actually *less* than its original coordinate along the axis given by the word *drive* on its own. Secondly, in the whole WORDSPACE there are several latent axes, including (hopefully) one for the general topic of computing, and the word *disk* will almost certainly have a much larger coordinate along this latent axis, which will dwarf its remaining affinity with the 'cars and driving' axis.

By using only coordinates along the k most significant latent axes (that is, those axes along which the variation between the data points is greatest), the number of dimensions is reduced from the number of original content-bearing words down to our chosen k. In smaller terms, if we replace the axes for *car* and *drive* with a single latent axis, we've replace two dimensions in the original model with a single dimension in the reduced model.

Some of the latent axes from the BNC WORDSPACE are shown in

Table 6.4. Some of the first few axes seem to represent very general ideas which nonetheless have a kind of underlying clarity to them (axes 1 and 2); some are clearly signifying specific topics such as medicine (axis 18) and some (especially further down the list) appear to be more bizarre than semantic (axis 77).

This approach can be thought of as a (very simplified) version of decomposing words into semantic primitives — basic elements from which word meanings are derived (Wierzbicka, 1996). In an ideal world, each of our latent axes should correspond to one of these primitive elements of meaning, and the coordinates a_i of a word vector a should tell us the extent to which each primitive element contributes to the meaning of the word a. We have indeed come a long way from the graph in Chapter 2, where everything was defined in terms of individual word–word relationships. We now have a set of fixed, global coordinate axes and every word is built up as a linear combination of these basic concepts.

The process of reducing the number of dimensions in a WORDSPACE to concentrate the information it contains is often called *latent semantic analysis* or LSA. It was first used for information retrieval by Deerwester et al. (1990), and it has been shown that LSA can sometimes help search engines because it can match queries and documents which share underlying topics even if they don't contain

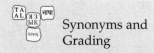

Synonyms and Grading

Latent semantic analysis has shown remarkable (possibly even disturbing) successes at some human language tasks. In a standard English synonym test (of the "The meaning of A is most like the meaning of B, C, D or E" multiple choice variety), LSA obtained an average score that would apparently have been sufficient for entry to many universities (Landauer and Dumais, 1997).

Even more surprising, comparing the document vectors of students' answers to essay questions (such as "Describe the operation of the CPU in a computer") to an 'ideal' answer has been shown to assign very similar grades to those given by professional examiners. Since the word *not* is usually ignored or has very little effect on an overall document vector, you should be able to trick such an algorithm into giving you an A grade by writing an essay containing all the relevant terminology and in which every sentence was a deliberate lie.

TABLE 6.4 The top neighbours of some of the latent axes from the BNC WORDSPACE

Axis 1		Axis 2	
back	0.413964	technique	0.342441
cried	0.410985	simple	0.340493
looked	0.401956	characteristics	0.332049
suddenly	0.381296	method	0.331636
quietly	0.373380	techniques	0.321635
laughing	0.367387	easily	0.317631
sighed	0.363614	limitations	0.310266
watched	0.362208	readily	0.308681
grinned	0.359385	linear	0.303569
silently	0.358613	processes	0.302632
laughed	0.358193	concepts	0.300701
walked	0.355388	necessarily	0.300475
afraid	0.354490	patterns	0.298793
stared	0.352614	measurement	0.298586
glanced	0.351504	occur	0.296558

Axis 18		Axis 77	
crohn's	0.375452	obscurity	0.415469
biliary	0.363433	alkali	0.410941
bladder	0.354028	brink	0.402437
colorectal	0.353670	detriment	0.401960
chronic	0.351685	cyclic	0.399894
gastrointestinal	0.344054	flicking	0.397805
bowel	0.342489	levers	0.394036
cardiac	0.339265	needles	0.364760
disease	0.337526	tara	0.359435
liver	0.335442	maids	0.358849
crises	0.333458	transferable	0.356399
gall	0.333248	qa	0.351020
biopsy	0.329915	filename	0.349595
inflammatory	0.329057	inlet	0.348850
tumour	0.327842	orcs	0.346828

exactly the same keywords (Berry et al., 1995). Like the vector model itself, LSA has grown out of its beginnings in information retrieval to contribute to other linguistic tasks such as word sense disambiguation (Schütze, 1998). The idea that we might combine information from different sources and contexts by recognizing underlying dependencies between superficially different variables has been suggested as an answer to the psychological question "How do humans manage to learn so much about language and the world in such a short lifetime?" which can be traced to Plato (Landauer and Dumais, 1997).

6.3 Clustering related words into concept groups

The diagrams of words related to *fire* in Tables 6.1 and 6.2 are typical of the way in which most search engines still present their results. The similarity or relevance score between a word (or more normally a document) and the user's query is computed for each candidate, and a list of the top few results is printed out as a sort of 'relevance hi-score table.' This can be desirable for a specific search for precise information, but it is less well adapted for exploring a wider range of information or for modelling word meanings in general.

A geographical analogy helps to illustrate this point. Suppose you wake up one morning to find yourself in strange surroundings, in an unfamiliar city. Naturally, you want to find out about your new location, so you ask a passer by "What things are there around here?" A traditional search engine might give an answer very like

- There is an Italian restaurant 127m away.
- There is a bus station 421m away.
- There is a river 1043m away.
- There is a lake 1167m away.

and so on. (In practice, most search engines don't even print out these distances, and just return a list saying "the nearest document is ... ," "the next nearest document is ... ," and so on.)

For information retrieval, this is often fine. You already know what sort of information you're looking for, and the search results allow you to narrow in for more details. For representing and understanding new concepts, however, it's a very bad way of giving information. If you're trying to understand the general shape of your new surroundings, you

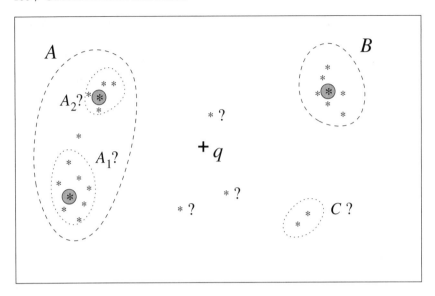

FIGURE 6.3 A collection of points and some of the clusters they may be partitioned into. The points in circles are the most central or prototypical points in their respective clusters.

don't just want to know what's nearby: you want to know how things relate to one another. For example, the river and the lake are virtually the same distance away: you might guess that they're related in some way. Does the river in fact flow into the lake? This leaves you asking not only how far the river and the lake are from where you are standing, but how close they are to one another.

One way to get a good idea of the lie of the land in WORDSPACE is to use *clustering*. The purpose of clustering is to divide a collection of points or observations into a discrete collection of different classes or *clusters*. A pictorial example is given in Figure 6.3, which shows the potential benefits and some of the difficulties of clustering. There are two clear clusters, one on the left marked A and one on the right marked B. However, there are a number of points that don't fit into either of these clusters — should these be regarded as just anomalies or should some of them be considered as separate clusters in their own right (such as C)? The A cluster on the left isn't altogether simple, either — it seems to have enough internal structure that we might choose to regard it as really consisting of two subclusters A_1 and A_2.

TABLE 6.5 Neighbours of *plant* (BNC data) and *fire* (NYT data) divided into different clusters

Prototypical example	Cluster members
plants	plant plants abundant seeds fertiliser fungi algae aquatic containers poisonous vine temperate habitat arctic herbs ponds grow flora fruits insect medicinal pests
radioactive	chernobyl reactor reactors generating radioactive nuclear sellafield plutonium reprocessing uranium mw radiation greenpeace bnfl disposal radioactivity cegb waste environmentalists stations hydro scientist power sizewell
chemicals	chemicals toxic chemical organic contamination bacteria pesticides sewage hazards ici
dioxide	wastes biomass dioxide nitrogen greenhouse fuels sulphur fuel electricity carbon polluting emissions conserve fossil oxide pollutants
foliage	planting flowering flower buds planted foliage flowers evergreen seed shrubs bulbs winter stems seedlings compost spring
factory	machinery robots factory factories
crops	soil fertilisers nutrients vegetation crop harvested cultivation crops

Prototypical example	Cluster members
fire	fire firefighters blaze fires flames brush marine smoke confrontation deadly spraying firing tense fired broke hose dump officials danger ants threatening burns
cease	cease ira truce paramilitary fein loyalists
gunfire	gunfire wounded sniper grenades raids clashes raid gunmen attackers unarmed violently gunman alert incident dead uniformed alerted engulfed wounding erupted police bullet bullets patrol ambulance policemen
fumes	trapped fumes burning sprayed tear leak accident accidentally monoxide alarm hazard detectors
rubble	burned rescuers evacuated rubble crews destroyed siege rescue corpses debris smoldering charred damaged rescued
capt	firefighter lt sgt capt maj accidental spokesman dispatched officer's
artillery	evacuation mortar civilians casualties hostage sarajevo helicopter convoy artillery soldiers tanks

In spite of these remaining questions, it still seems clear that if these points represent words or documents that are near to a query represented by the point q, then an algorithm that could say "some of the results are in an A group, some are in a B group and some are in neither" would be telling us more about the data than one that just listed the results ordered by their distance from q (which would mix up the A and B groups and leave the user to try and work out which is which). When exploring a WORDSPACE, clustering enables humans to see the principal categories into which words can be grouped, and to gain a lot of information about the different concepts around which a local region of the WORDSPACE is built, at a single glance.

Two examples are given in Table 6.5, one showing the top 100 neighbours of the word *plant* in the BNC, and one showing the top 100 neighbours of the word *fire* in the NYT corpus. Each dataset has been divided into seven clusters, and these clusters certainly seem to encapsulate some of the different senses of the words in question. For example, some of the *fire* clusters are clearly related to the military sense of *fire* and some are related to the civilian sense: some of the *plant* clusters are related to the living-thing sense of *plant* and some are related to the factory and production sense. The "Prototypical Example" in the left hand column is the word which is most central in each cluster (the points we've circled for clusters A_1, A_2 and B in Figure 6.3).

The algorithm we used for these examples was motivated to some extent by the ideas of Prototype Theory (see Chapter 4), in that we tried to find good example words for different regions and assigned the other words to clusters based upon which example word they were closest to (the algorithm is given in full in Table 6.6). This particular clustering algorithm has its drawbacks: for example, which points become chosen as prototypes depends on which have already been chosen, and so the first one chosen right at the beginning is often simply an average of all the data being clustered.

The main point of this section is that once you have a collection of word vectors and a way of measuring distances between them, you can group those vectors into clusters which represent different uses, different contexts, and often different *senses* of words. There are many different clustering algorithms which can be used for this purpose (Manning and Schütze, 1999, Ch 14), and it has become one of the most

TABLE 6.6 **Prototype generating algorithm.**

Let $d(a, b)$ be a metric on WORDSPACE (we used the Euclidean metric) and let W be a set of word-vectors. Each word-vector $w \in W$ will be assigned a pointer to a prototype word $p_w \in W$ in whose cluster w belongs. These pointers are initially null, and the distance from w to p_w is initially large.

1. For all words $w_i \in W$,

2.   for all $w_j \in W$, calculate $d(w_i, w_j)$

3.   if $d(w_i, w_j) < d(w_j, p_{w_j})$

4.   record a gain of $d(w_j, p_{w_j}) - d(w_i, w_j)$

5.   the sum of all these gains is the total that would be gained by adding w_i as a new prototype point

6. Declare the word w_{best} with the biggest gain to be a new prototype point

7. For all words w_j which are closer to w_{best} than to their previous prototype point, assign $p_{w_j} = w_{\text{best}}$.

8. Repeat from step 1 until the desired number of prototypes is found.

---

recognized techniques for learning about ambiguous words from their occurrences in corpora (Schütze, 1998, Lin, 1998a).

6.4 Plotting Word-Senses in two dimensions

Demo: http://infomap.stanford.edu/visual

Clustering can be a good way to learn about which words are similar to one another in WORDSPACE, but sometimes we really want to see this information graphically. Going back to the analogy of waking up in a strange town and trying to find our way around — we might have been told which objects lie in clusters together, but suppose we really want to see a *map* of this information. For geographical maps, this is no problem: points on the earth's surface only need two coordinates to determine them (for example, latitude and longitude), so for any local area, projecting this surface onto a flat screen or a piece of paper is simple and effective.

Things are more complicated in WORDSPACE. Even after we've used Latent Semantic Analysis and projected down to (say) 100 dimensions, we still have far too many words and too many dimensions to visualize effectively. Both of these problems can be solved. First of all, we restrict our attention to a few related words in a particular region of WORDSPACE, and then we plot these words by reducing everything to just *two* coordinates which can be plotted on a screen.

The first of these steps is simple: we find the nearest neighbours of a word or query just as in Table 6.1, and we select these words and their vectors for a closer analysis. We then use exactly the same algorithm that projects onto our 100 significant dimensions in WORDSPACE to project onto the two coordinate axes which give the best latent axes for this reduced data set. (This is described in more detail by Widdows et al. (2002a).) This is really a higher dimensional analogy of the familiar process of finding a best-fit line for a collection of points in two dimensions (as shown in Figure 6.2). The only difference is that we're finding a best-fit *plane* for a collection of points in 100 dimensions.

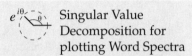

Singular Value Decomposition for plotting Word Spectra

We obtain our two coordinate axes for plotting a word spectrum by using the singular value decomposition (page 175) on the reduced data set of word vectors in a particular region of WORDSPACE. The top rows of the matrix V^* give the most significant axes: the leftmost columns of the matrix $\hat{U}$ give the coordinates of each point along these axes.

In practice, the first axis seems mainly to say "most of the data's over here," and doesn't tell us much about the variations *between* data points: because of this, we've found that plotting the second and third columns of the matrix $\hat{U}$ gives much better results than plotting the first and second columns. These coordinates are then rescaled so that the words fit nicely on the screen.

We called the resulting pictures *Word Spectra* because they try to give a spectrum of the different groups of words and meanings associated with a target word. You can generate them online by visiting the URL at the beginning of this section or by clicking on the "Visualize these Results" link on the main WORDSPACE demo (Figure 6.1). Two examples are given in Figure 6.4, showing Word Spectra of the words *drug* and *word*, built from the NYT WORDSPACE (and slightly edited by the author so that fewer words land right on top of one another, which is a hazard when we project from many dimensions down to just two). The *word* picture has a very distinct cluster in the top left of the image, corresponding to the computer program *Microsoft Word*, and the spectrum accurately represents the way in which this meaning of *word* is separate from all the others. The different meanings of *drug* are also well-represented in Figure 6.4, with the illegal drugs being together on the left hand side, words about medical conditions and treatments being at the bottom right, and a cluster of drug and pharmaceutical companies being together at the top.[4]

Pictures like this can be a very good way of understanding why different words come out as being similar in different models: you can often get a very clear picture of how different words are used in different kinds of writing. This highlights interesting contrasts and tendencies that you might not immediately have expected but which gradually make sense when you see them in the context of other related words.

6.5 Mapping between two different WORDSPACES

So far in this book, all of our models have been built from a single corpus, and keeping them separate has certain benefits, such as providing good models of the way concepts are used in particular domains. If we combine models by simply concatenating the corpora they are built from, much of this distinction would be lost. For example, the word *operation* behaves very differently in our medical WORDSPACE built from the Ohsumed corpus (where the neighbours of *operation* include *reoperation, corrective* and *surgery*) than it does in our financial model built from the Wall Street Journal (where the neighbours of *operation* include *expand, streamline* and *unprofitable*). It would clearly be very foolish to assume that a word like *operation* can be transferred in between models without carefully checking to see if the meanings are consistent.

One way to check for this consistency is to translate not only an individual word vector, but a collection of related vectors. We can accomplish this by considering the neighbours of a particular word in

---

[4]The shape of this diagram has a striking resemblance with the British Isles. The hearty Irish *craic* has been mistaken for something completely different; the Scots are very business oriented (*upjohn* being Newspeak for John'o'Groats); after many centuries Her Majesty's Customs Officers are still trying to crack down on smuggling in Cornwall; they're having a terrible time on the London Underground again: but at least there's *progress* in West Yorkshire.

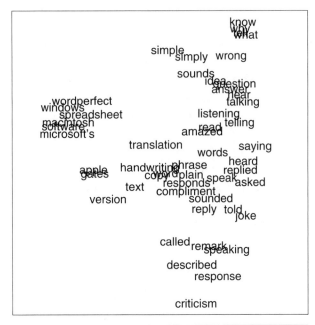
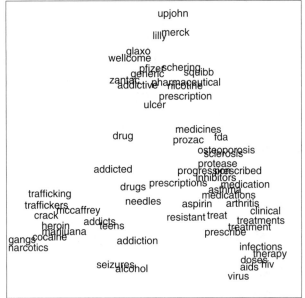

FIGURE 6.4 Word Spectra for the regions surrounding the words *word* and *drug* in a WORDSPACE built from NYT data

one WORDSPACE, and try and find the place in another WORDSPACE where these neighbours are concentrated (rather like the class labelling technique for mapping a list of words into a taxonomy introduced in Section 3.6).

For example, in the WORDSPACE built from the BNC, the word *labour* has neighbours including *party, liberal, leader* and *elections*, because the most frequent way the word *labour* is used in the BNC is as a term for the political **Labour Party**. From the New York Times, the word *labor* has neighbours which include *wages, productivity*, and *strike* — totally different from the BNC *labour*. On the other hand, the word *democratic* in the NYT has neighbours that also include *party, liberal, leader*, and *elections*.

We might infer from this that the predominant meaning of *labour* in the BNC is much closer to the meaning of *democratic* in the NYT than to the meaning of *labor*. This would be quite correct — the predominant meaning of *labour* in the BNC is the name of a political party, just like *democratic* in the NYT.

To turn this intuition into an algorithm, we proceed as follows. Suppose we have two different WORDSPACE models S and T (think Source and Target), and we wish to map the vector v from S to an appropriate vector in T. Now, suppose that the neighbours of v in the source model S include the vectors a, b and c, and that these words also have vectors $\tilde{a}$, $\tilde{b}$ and $\tilde{c}$ in the target model T. We measure the distances $d(v,a)$, $d(v,b)$ and $d(v,c)$ in the source model S: we then try to find the vector $\tilde{v}$ in the target model T so that the new distances in $d(\tilde{v},\tilde{a})$, $d(\tilde{v},\tilde{b})$ and $d(\tilde{v},\tilde{c})$ match the old distances $d(v,a)$, $d(v,b)$ and $d(v,c)$ as closely as possible. This process is depicted in Figure 6.5.[5]

Table 6.7 shows the results of using such a mapping algorithm to transfer semantic information between models built from British and American English. The mapping correctly predicts that a speaker of British English looking for articles about *football* in the USA should look for the word *soccer*, and conversely, an American seeking to buy *gas* for their car in the UK should ask for *petroleum*, not *gas*.

Though these examples are somewhat playful, there is a serious side

---

[5] The analytic details are rather more complex and there are many ways of writing equations to give suitable mappings (varying according to how many neighbouring points are used, how the possible redundancies among these neighbouring points should be accommodated, how much distortion is allowed for each neighbouring point, and so on).

TABLE 6.7 The results of transferring individual words between models from different kinds of English, and the better translations obtained using a semantic mapping based upon the word's neighbours.

BNC		WSJ		Mapped from BNC to WSJ	
football	1.000000	football	1.000000	soccer	0.715946
soccer	0.789099	basketball	0.917423	clubs	0.673636
players	0.719179	league	0.897629	fans	0.619927
league	0.715559	baseball	0.882225	players	0.603639
fans	0.687709	championship	0.870277	league	0.590624
clubs	0.684717	hockey	0.869916	baseball's	0.590203
player	0.680154	teams	0.822637	tournament	0.584194
game	0.678265	league's	0.821501	player	0.584172
united's	0.672423	nba	0.816337	sport's	0.578925
rugby	0.671691	playoffs	0.816257	cup	0.568693
fa	0.670602	fans	0.801249	baseball	0.568642

The semantic mapping approach does a good job of translating *football* in the British National Corpus model to *soccer* in the Wall Street Journal model, where the word *football* means a different sport.

NYT		BNC		Mapped from NYT to BNC	
gas	1.000000	gas	1.000000	petroleum	0.738208
natural	0.900443	coal	0.727908	exploration	0.712915
pipeline	0.817075	electricity	0.694651	drilling	0.689073
pipelines	0.762193	fuels	0.687560	oil	0.662095
petroleum	0.751568	oil	0.664746	fuels	0.593865
oil	0.734236	petroleum	0.583274	offshore	0.549640
fuels	0.725791	fuel	0.572129	mining	0.519150
rigs	0.721321	heating	0.530885	pipeline	0.506894
exploration	0.720229	heat	0.527546	natural	0.479182
drilling	0.705463	energy	0.516645	iaea	0.464378
gasoline	0.676359	electric	0.516234	coal	0.462963

The semantic mapping correctly translates *gas* from the New York Times model to *petroleum* in the British National Corpus model.

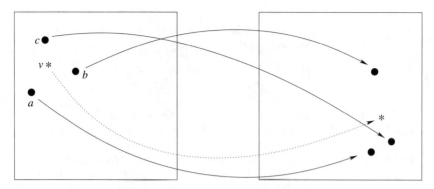

FIGURE 6.5 A sketch of the process of mapping the vector v from one model to another by preserving its position relative to its nearest neighbours.

to this work. The goal is to find ways of helping people and natural language processing systems to recognize when a word is being used with a familiar meaning, and when on the other hand, it is being used with a completely different meaning from what they are used to. Ideally we can even go one step further and suggest alternative terms that *can* be used to convey the meaning intended.

6.6 Bilingual vector models

Demo: http://infomap.stanford.edu/bilingual

One practical situation where a person may know very few of the meanings associated with the words in a particular document arises when the documents are written in an unfamiliar language. In this case, the neighbouring words of a target word w may be of little help, because the user won't be familiar with these words either. This section describes how a WORDSPACE built from translated documents can help in such a situation.

If we have a collection of documents that are translated into two languages, it's actually possible to build a single WORDSPACE which represents words in *both* languages (a technique that can also be used to build a search engine which finds documents in both languages (Littman et al., 1998, Yang et al., 1998)). Such a translated corpus is called a *parallel corpus*, and apart from widely translated works of literature (which are comparatively short by the tens of millions of words standard), many of the most useful parallel corpora have arisen for political reasons: for

TABLE 6.8 An example of a compound document, a German original paired up with its English translation.

Anatomie des hinteren Kreuzbandes	Anatomy of the posterior cruciate ligament
Das hintere Kreuzband ist das kraeftigste Band des menschlichen Kniegelenks. Es entspringt faecherfoermig an der Innenflaeche des Condylus femoris medialis und inseriert im hinteren Anteil der Area intercondylaris. Es besteht aus einer Vielzahl kleiner Faserbuendel. Unter funktionellen Gesichtspunkten koennen femoral anterior entspringende Fasern von femoral posterior entspringenden Fasern unter schieden werden.	The posterior cruciate ligament (PCL) is the strongest ligament of the human knee joint. Its origin is at the lateral wall of the medial femoral condyle and the insertion is located in the posterior part of the intercondylar area. The posterior cruciate ligament consists of multiple small fiber bundles. From a functional point of view, one can differentiate fiber bundles with an anterior origin and fiber bundles with a posterior origin at the femur.

example, the proceedings of the Canadian parliament are recorded in both English and French, and versions of many official European Union documents are produced in not just two but several languages.

The main idea in building a bilingual WORDSPACE is to consider each document along with its translation as if they were a single 'compound' document. An example of such a compound document is given in Table 6.8. Two words that occur in documents which are translations of one another can then be regarded as cooccurring, just as if they had occurred in the same monolingual document (Figure 6.6).[6] The corpus we used in our experiments was a collection of abstracts from German medical documents which are available with English translations.[7]

This model allows a user to enter terms in *either* language and find related words or relevant documents in *both* languages. An example

---

[6] If documents are long and contain various different topics, it's important to *align* the bilingual text so that you know which smaller units such as sentences are translations of one another (Melamed, 2000). Fortunately, a corpus of short, single-topic documents such as the journal abstracts we used in our experiments is naturally pretty well aligned.

[7] These abstracts were made available by the Springer Link website (link.springer.de) and collected into a corpus by the MuchMore project (muchmore.dkfi.de) for the development of a concept-based medical information system that can work in both English and German.

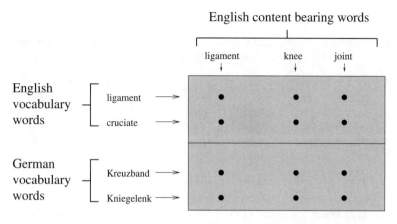

FIGURE 6.6 Measuring cooccurrences between terms in a compound English–German document

using the English query *bone* is given in Table 6.9 — the bilingual WORDSPACE correctly picks out the German word *knochen* as the closest counterpart of the English word *bone*. Since the bilingual WORDSPACE seemed to be doing a good job of mapping English terms to their German counterparts, we evaluated it to see how reliable it is by comparing the translations given by the bilingual WORDSPACE with English–German term pairs given by the UMLS multilingual medical thesaurus. We found that for words with high cosine similarity scores, their nearest neighbour in the other language was an accurate translation over 90% of the time. This technique even found German translations that hadn't made it into the UMLS medical thesaurus yet, in particular acronyms that were listed only as English terms but which are also used in German with the same meaning. These experiments (and the technique for building a bilingual WORDSPACE from a parallel corpus) are described in much more detail by Widdows et al. (2002b).

The assumption that each English word should be translated to a single German word is of course very naïve, and was responsible for many of our errors in these experiments. Many German words are compounds corresponding to several English words rather than just one. To some extent, the bilingual WORDSPACE model can model this phenomenon: for example, Table 6.10 shows that the closest German neighbour of the English query *lung transplant* (formed by adding

TABLE 6.9 Words in both English and German which are near to the English query *bone* in the bilingual WORDSPACE

English neighbours		German neighbours	
bone	1.000000	knochen	0.823083
cancellous	0.700623	knochens	0.708817
osteoinductive	0.671816	knochenneubildung	0.699606
demineralized	0.648947	spongiosa	0.635176
trabeculae	0.639279	knochenresorption	0.595616
formation	0.595301	allogenen	0.594648
periosteum	0.562293	knöcherne	0.590172
osteoporotic	0.561281	knochenheilung	0.578918
autoclaved	0.559798	bone	0.569451
augmentation	0.543297	knochentransplantate	0.565430
substitute	0.532057	knochentransplantaten	0.564502
hydroxyapatite	0.528326	trabekulären	0.555980
ridge	0.526757	knochentransplantation	0.548806
osteoclast	0.523437	aufgefüllt	0.545810
marrow	0.523071	hydroxylapatitkeramik	0.542906
resorption	0.516087	knochenregeneration	0.531353

TABLE 6.10 Words in both English and German which are near to the English query *lung transplant* in the bilingual WORDSPACE

English neighbours		German neighbours	
transplant	0.794835	lungentransplantation	0.730531
lung	0.794835	lunge	0.649870
transplantation	0.608974	transplantiert	0.520466
bronchiolitis	0.586815	lungenemphysem	0.516735
recipients	0.586446	transplantation	0.510346
obliterans	0.545399	lungen	0.510102
actuarial	0.538949	transplantatvaskulopathie	0.496594
rejection	0.521160	lungenfunktion	0.483037
lungs	0.513533	plötzlich	0.466997
pneumonectomy	0.497044	htx	0.458539
orthotopic	0.492516	alveolen	0.442872
allograft	0.477084	pneumonektomie	0.438988
vasculopathy	0.475079	organtransplantation	0.438770
donor	0.472244	ards	0.435389
transplantations	0.460281	lungenerkrankungen	0.434309
bronchial	0.445858	pulmonalen	0.433340

the English *lung* and *transplant* vectors together) is the compound term *lungentransplantation*. This example works particularly well partly because the meaning of *lung transplant* is *compositional*, meaning that it is simply deduced from the meanings of its constituent parts. There are many compound phrases that are not compositional (such as *Big Apple* and *red herring*), and for these phrases simply adding together the vectors of the individual words gives a very poor representation of the idiomatic meaning of the compound term. (We have had some preliminary success in recognizing such non-compositional phrases precisely by looking for those phrases whose distributions are different from those of their constituent words; see Baldwin et al., 2003).

The Word Spectrum plotting technique of Section 6.4 can be particularly effective for visualizing words in more than one language, which can help a user to relate words in an unfamiliar language to words in their preferred language. An example output for the English query word *drug* is shown in Figure 6.7 (in the online version,[8] English words appear in red and German words appear in blue: the method of adapting this to black-and-white by putting the German words into gothic script is derived from Goscinny and Uderzo, 1974).

As we saw in Figure 6.4, the word *drug* in English can be used with quite different meanings, and in German these two meanings are actually represented by different words (*Medikamente* = prescription drug and *Droge* = narcotic). To correctly translate an English sentence containing the word *drug* into German, it is therefore not enough to look up the German word for *drug* in an English–German dictionary: it is important to know which *meaning* of *drug* is being used so that you can decide which of the possible German translations in correct. The two-dimensional plot in Figure 6.7 shows these two different areas of meaning quite effectively, and a user faced with the two possible translations of the word *drug* could easily use this picture to work out which meaning corresponds to *Droge* and which meaning corresponds to *Medikamente*. Techniques such as this can be particularly valuable in multilingual applications, because they enable a user to connect with unfamiliar terms in a foreign language by relating them to known terms in a familiar language.

---

[8] `infomap.stanford.edu/visual`

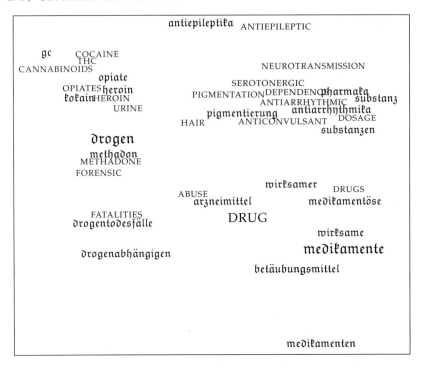

FIGURE 6.7 ENGLISH and German terms related to the English word *drug* in the Springer medical abstracts.

6.7 Using WORDSPACE and class labelling to enrich a taxonomy

Combining information from different models can be even more useful than building the models in the first place: one of the applications of WORDSPACE has been to supplement the information gathered in more traditional formats such as the WordNet noun taxonomy. This section briefly explains how this has been done by combining two of the tools we have introduced already, namely vector models and class labelling.

Nearly every corpus of English text will contain some words that are in WordNet, and some that aren't. The basic idea is to build a WORDSPACE so that for each word w in the corpus that *isn't* in WordNet, we can gather a collection of neighbours $N(w)$, at least some of which *are* in WordNet. We can then use the set $N(w)$ as the input to a class

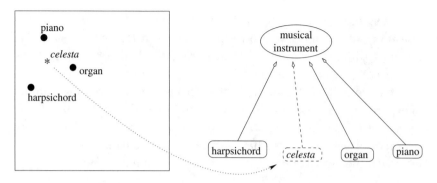

FIGURE 6.8 The neighbours of *celesta* in WORDSPACE can suggest where this word should go in a taxonomy

labelling algorithm, the hypothesis being that good class labels for the set $N(w)$ will also be hypernyms of the unknown word w.

This outline is, on the face of it, quite realistic. Suppose, for example, that you encountered the word *celesta*, not knowing that it referred to a musical instrument. If you heard the word *celesta* along with the words *piano, harpsichord* and *organ*, you might be tempted to infer that, since these other three are *musical instruments*, a *celesta* is a *musical instrument* as well (Figure 6.8).

The first successful algorithm along these lines was developed by Hearst and Schütze (1993), who partitioned the WordNet taxonomy into a number of disjoint sets which are used as class labels. This simplifies the taxonomy into just two levels: one level of species and one of genera, each of the specific words having a single hypernym or genus. A collection of words could then be assigned a best class label by majority voting, which provides a very simple version of the class labelling algorithm of Section 3.6 in which the affinity-score α between a word w and a class h is simply the set membership function, $\alpha(w, h) = 1$ if $w \sqsubseteq h$ and 0 otherwise. For each test word w, a set of neighbours was gathered using a WORDSPACE model very much like the ones described in this chapter, and this set was passed to the majority-voting class labelling algorithm, giving one of the first successful algorithms for mapping new words into taxonomies.

A direct descendant of this approach is described in Widdows (2003c), which uses the class labelling algorithm shown in Figure 3.8 (page 91) to map words into WordNet. This algorithm is able to choose for itself a suitable level of specificity or generality for the input words in question. This work also used a part-of-speech tagger to distinguish between different grammatical categories, so that (for example) the WORDSPACE distinguished between the noun and the verb uses of *fire*, which have quite different meanings. This is one of a growing number of experiments to have combined different forms of *linguistic* knowledge into a *mathematical* model. Results varied according to how many neighbours were used, whether parts of speech were distinguished, and how well the WordNet taxonomy represented the neighbours to begin with.

One of the principal difficulties of carrying out and evaluating work of this kind is knowing when to *contradict* a resource like WordNet. Though hand-built by several hours of diligent human labour, it is always a mistake to think that lexical resources contain the whole truth about a word, partly because ambiguity is so pervasive and systematic. A simple example illustrates this point: WordNet lists *Hampshire* as a kind of *sheep*, which of course it is. But *Hampshire* is also an English county, after which the sheep is named. (Many agricultural terms are derived from the names of geographical regions.) Our WORDSPACE model built from the BNC correctly associates *Hampshire* with neighbours including *Yorkshire, Lancashire, Derbyshire, Durham*, which are also English counties (and more specifically, counties with cricket teams). The neighbours of *Hampshire* in our WORDSPACE and the coordinate terms of *Hampshire* in WordNet are thus very different. This is not because either model is wrong, but because the two models have correct information about different *senses* of *Hampshire*. Ambiguity is ubiquitous but often unpredictable, and one of the things that makes working with language so difficult and so interesting.

Conclusion

This has been an all too brief guide to some of the possibilities for exploring concepts in WORDSPACE. There are many more which we haven't mentioned yet, and techniques for mapping in between WORDSPACE and some of the other models such as graphs and hierarchies. In many ways, we are still experimenting and feeling our way around these spaces, and there is a wealth of research yet to be done in learning which methods are particularly suited to different aspects of language.

Some of these mappings and combinations are suggested in the Wider Reading, and the next chapter will take us much further into exploring the possibilities for using *logical* operators to find important concepts in WORDSPACE.

Wider Reading

The past 10 or 15 years have seen a great deal of interesting work in representing words and concepts (and data from all sorts of other fields as well as linguistics) as vectors. The WORDSPACE we use in this book is based directly on the work of Hinrich Schütze (1997, 1998), who used vector methods to address problems in ambiguity learning and resolution, and for adapting lexical resources for specific applications (Hearst and Schütze, 1993).

Dimensionality reduction is an important area of research for many reasons: as well as natural language, such techniques can be important for analyzing biomedical and genetic data, image compression and reconstruction, and many other statistical problems where a few particularly important factors influence many many variables. Latent semantic analysis (Deerwester et al., 1990, Landauer and Dumais, 1997) is still the most widely used algorithm for these purposes in natural language processing. An alternative is *probabilistic* latent semantic analysis (PLSA), which pictures the latent variables as generators for probability distributions rather than as geometric axes (Hofmann, 1999), a technique that has been successfully used for text-segmentation by Brants et al. (2002). There are also some interesting *nonlinear* algorithms (Tenenbaum et al., 2000, Roweis and Saul, 2000) that are able to model curved geometric structures, whereas traditional singular value decomposition (the algorithm behind LSA) makes the assumption that the data is basically flat.

An overview of several clustering algorithms and their applications in natural language processing is given by Manning and Schütze (1999, Ch 14), and other works including Pereira et al. (1993) and Li and Abe (1998b). A collection of papers on using the expectation-maximization algorithm for building lexicons is available in Rooth et al. (1999), which includes bilateral clustering on verb–object pairs. The homepage of

Dekang Lin[9] contains an excellent online example of the use of similarity and clustering to obtain different word senses (Lin, 1998a, Lin and Pantel, 2002). Lin's algorithm automatically chooses how many clusters to return for each word, and where available provides good class labels for the clusters. Clustering has recently received a boost in prominence in information retrieval, with the launch of Vivisimo's *Clusty* search engine.[10]

The topic of visualization in information retrieval, and its role as part of a user interface, is discussed by Hearst (1999). The collection of papers in Card et al. (1999) provides an excellent overview of information visualization: it contains chapters on trees and networks; the problem of reducing information from many dimensions down to two or three; the opportunities provided by alternative geometries to focus on a particular region while keeping an eye on the bigger picture; and a whole section devoted to visualization for natural language documents.[11]

Related to the problem of mapping words into a taxonomy (as in Hearst and Schütze (1993), Widdows (2003c)), there has been considerable work on automatic taxonomic classification of *documents* (Dumais and Chen, 2000, Chakrabarti et al., 1998), an application which is of considerable interest to many web portal and search engine companies.

This book has by now introduced several different geometric spaces that can be used to describe words and their meanings, and we began in Sections 6.5 and 6.7 to consider the interesting possibility of mapping *between* these structures. There are some well-known transformations that accomplish precisely this purpose: for example, the nodes in a graph can easily be represented as points in vector space using the *adjacency matrix* of a graph (Bollobás, 1998, §II.3). This correspondence has many desirable properties — for example, a graph of symmetric relationships gives rise to a symmetric adjacency matrix (one which stays the same if you flip it across the main top-left to bottom-right diagonal). Vectors

---

[9] www.cs.ualberta.ca/~lindek/demos/wordcluster.htm

[10] www.clusty.com

[11] A recent survey of commercially available information visualization software can be found in *The Economist Technology Quarterly* (June 2003, pp. 25-26). At the time of writing, it seems possible that information visualization technology may transform the whole process of browsing the internet: but it is also perfectly possible that it might not.

of this sort are used by Caraballo (1999) in work which combines several lexicosyntactic patterns with hierarchical clustering to build a concept hierarchy automatically. Cederberg and Widdows (2003) further combine lexicosyntactic patterns with graph-based similarity and WORDSPACE similarity, showing that the different mathematical models can successfully act as filters and sanity-checks for one another's results, considerably improving overall accuracy.

The transformation from a graph to a vector space can be inverted by interpreting the cosine similarity of two words a and b to be a measure of the strength of the link between a and b in a weighted graph — a simple unweighted graph can then be obtained by considering two words as being linked if this similarity exceeds a given threshold λ (so that $a \leftrightarrow b$ in the graph if $\cos(a, b) > \lambda$ in the WORDSPACE).

7

Logic with Vectors: Ambiguous Words and Quantum States

As we try to learn and reason about language and its meaning by measuring the distributions of points and regions in geometric models, a variety of powerful logical operators become available. Many such operators correspond to well-known logic gates in computer science, and to familiar options in standard user interfaces.

A typical example is shown in Figure 7.1. In many browsers and user interfaces (in the year 2004), if you click on one icon and then click on another, the second will become selected and this will cancel the selection of the first, and if you click on more icons, the one most recently selected becomes the unique active icon, overiding previous selections. However, if you hold down the CTRL key and then click on two different icons, *both* icons will be selected. And if you hold down the SHIFT key and then click, *both* icons *and those in between* will be selected. When using these features, previous selections are not discarded — all the icons selected while holding down the correct key are combined to give a special set of selected items. The processes of combination correspond to *disjunctions* (*OR* operators or *joins*) in different logics.

The 'CTRL + click' option is essentially a discrete approach which builds a set whose elements are two isolated points. The 'SHIFT + click' option is a more continuous approach, which uses the geometry of the surrounding region, building a set consisting of all points contained in the rectangle spanned by the two icons the user selected.[1]

---

[1] This rectangle contains all points that lie on a shortest path or *geodesic* between the two generating points, where distance is measured using the Manhattan metric (page 102). By

Logic with Vectors: Ambiguous Words and Quantum States / 201

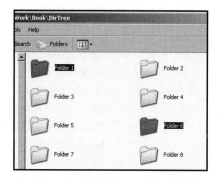 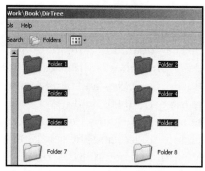

FIGURE 7.1 Holding down CTRL and clicking on *Folder 1* and *Folder 6* highlights these two folders only (left). Making the same two clicks while pressing SHIFT highlights Folders 2 to 5 as well (right). The first version is a *Boolean disjunction*, the second is a kind of *quantum disjunction*.

This chapter describes these two brands of logic, using the algebra of George Boole as the classical example of a discrete (and indeed, binary) approach to logic, and the quantum logic of Garrett Birkhoff and John von Neumann as the standard example of a continuous logic. Geometrically, the differences between these approaches arise from the differences between set theory and vector spaces, which form the underlying spatial models for Boolean logic and quantum logic respectively.

Because vector spaces provide the spatial model for both quantum logic and the Infomap WORDSPACE, it turns out that we can use the same logical operators in both. For me, this story began when I implemented a vector negation operator in WORDSPACE using the concept of orthogonality, just as described by Birkhoff and von Neumann in 1936. This precedent was happily brought to my attention by Professor Lawrence Moss of Indiana University, as the research evolved.

Exploring such operations, comparing their effects on document retrieval, and piecing together the scientific context of the WORDSPACE model proved an interesting journey. Many historic branches of logic, geometry, and the physical sciences became woven into the story, and

---

using different metrics, similar forms of spatial disjunction can be implemented in many other spaces.

I shall explain how these traditions have been used in search engines, especially in WORDSPACE.

The 'quantum' thread in this chapter stretches from Aristotle's description in the *Metaphysics*, through geometric implementations by Euclid, Descartes, and Grassmann, the birth of quantum mechanics itself in the early 1900s, and the outward stimulus this has provided to areas of logic, philosophy, computation, and word sense discrimination. The quantum idea, and the spatial extensions with which it is modelled, is a useful tool both for describing the way a particle may be represented by a combination of possible pure states, and for describing the way an ambiguous word may be represented by a combination of available 'pure' meanings.

First, for necessary context and contrast, we explore some of the traditional combination operations in discrete logical models, focussing on the work of George Boole.

7.1 Boolean Logic and Algebra

Boolean algebra is used in every computer, every pocket calculator, every microchip. The binary digit itself, and the logic which applies it to so many situations, were the brainchild of George Boole (1815–1864), a Lincolnshire schoolteacher whose family commitments prevented him from giving up his day job to attend University. However, Boole's efforts to teach himself mathematics were encouraged by Duncan Gregory, a fellow at Trinity College in Cambridge, and Boole began to publish his research in the *Cambridge Mathematical Journal*. His work became well-known and widely respected, and in 1849 Boole was appointed Professor of Mathematics at Queens College, Cork, in Southwest Ireland, where he continued his outstanding career as a dedicated scholar and teacher until his death.

The year 1854 saw the publication of Boole's masterpiece, *An Investigation of the Laws of Thought, on which are founded the Mathematical Theories of Logic and Prababilities* — one of the most influential works of modern mathematics, logic, and a milestone in the development of computer science almost a century before electronic computers were built.[2] Boole believed that his work directly concerned "the science of intellectual operations," stating that

> It is designed, in the first place, to investigate the fundamental laws of those operations of the mind by which reasoning is performed.
>
> (Boole, 1854, p. 3)

Both Aristotelian logic and probability are presented as applications: in contrast to the view that logic is reliable and certain and probability is messy and uncertain, Boole regards probability as having "presented far more of that character of steady growth which belongs to a science," his reasoning being that logic had been influenced by the to-and-fro of years of debate and fashion back through the middle ages to Aristotle, whereas probability had steadily matured since its invention by the French mathematician Blaise Pascal (1623–1662) in an attempt to curb an aristocratic friend's gambling addiction.

In spite of all these influences and goals, Boole's book is largely about *algebra* — replacing concepts with symbols, and then manipulating those symbols as if they were numbers, thus enabling different equations to represent different statements about concepts. For example, if the symbol x represents the class of **white objects** and the symbol y represents the class of **sheep**, the symbol xy can be used to represent the class of **white sheep**. Writing this the other way round as yx (**sheep which are white**) results in the same class, so $xy = yx$ and Boole's product operation is commutative. Equations with these symbols can then be interpreted as propositions — for example, if y is the class of **sheep**,

$$y = 0 \quad \text{means that} \quad \text{"There are no sheep."} \tag{7.13}$$

If you multiply zero by anything, the answer is still zero, so multiplying both sides of this equation by the symbol x gives the result

$$xy = 0 \quad \text{means that} \quad \text{"There are no white sheep."} \tag{7.14}$$

---

[2]Some of Boole's core ideas were published in his earlier book, *The Mathematical Analysis of Logic*, in 1847. This work is different in flavour from *The Laws of Thought*: it reads more like a set of mathematical lecture notes than a scientific treatise for the general reader. For a more thorough comparison, see the wider reading section at the end of this chapter.

In this way, Boole's binary algebra of classes can be used to draw the conclusion "There are no white sheep" from the premise "There are no sheep."

The classes described by Boole are exactly what we would call sets nowadays, and Boole's product operation on classes corresponds to their intersection as sets, which would be written in modern notation as $x \cap y$. Now, the intersection of any set x with itself is the same set x — the combination of the class of **white objects** with the class of **white objects** is still the class of **white objects**.[3] This standard identity is written $x \cap x = x$ in modern set theory, but Boole expresses the relationship using the algebraic equation

$$x^2 = x \qquad \text{or equivalently} \qquad x(1-x) = 0. \qquad (7.15)$$

This equation is true for all classes x, but the only *numbers* for which it is true are 0 and 1. For this reason, Boole (1854, Ch II) urges the following:

> Let us conceive, then, of an Algebra in which the symbols x, y, z, etc. admit indifferently of the values 0 and 1, and of these values alone.

Allowing for due poetic license, one might claim that in this declaration, the binary digit or 'bit' of modern computing was born. Boole goes on (Ch III) to equate the symbol 0 with *Nothing* (in modern terms, the empty set $\emptyset$) and the symbol 1 with *Universe* or *Everything*, and much later (Ch XI) to associate the equations $x = 1$ and $y = 0$ with the truth and falsehood of different propositions. In the light of the modern applications, it is interesting to see that the choice of the binary values 1 and 0 (which are called Boolean values to this day) was motivated first of all by the purely *algebraic* considerations of equation (7.15).

Just as Descartes had applied algebra to geometry 200 years earlier, Boole applied algebra to statements in language, translating dictionary definitions such as

> **Wealth** consists of things transferable, limited in supply, and either productive of pleasure or preventive of pain.

into equations like

$$w = ts(p + r), \qquad (7.16)$$

---

[3]This principle is often violated in natural language — for example, a **white, white sheep** is not quite the same as a **white sheep**, the repetition of **white** signifying that this sheep is *especially* white. This could be viewed as one of the many situations where *some* difference in meaning will develop between otherwise synonymous words or phrases, as speakers exploit small differences to enable greater expressivity.

where w = *wealth*, t = *transferable*, s = *limited in supply*, p = *productive of pleasure* and r = *preventive of pain*. For any object x in a universal set U, given its values for the variables t, s, p and r, the equation in (7.16) can be used to determine whether or not x can be described as *wealth*. With this, and other wonderfully Victorian examples such as

> Clean beasts are those which both divide the hoof and chew the cud.
> (*Deuteronomy*, 14:6)

the reader is taught an algebraic system which Boole then uses to express symbolically the arguments of great philosophers including Clarke, Spinoza and Aristotle.

Referring to the variables such as w and t as constituents (basis vectors) and the values of these variables as coefficients (coordinates) enables the object x to be represented by its coordinates along the t, s, p, and r axes — a vector, like the currency vectors in Chapter 5, only this time the coordinates, instead of being any real numbers, can take only the *Boolean* values 0 and 1.

Boolean-valued (i.e binary-valued) arithmetic lies at the heart of computing. Tiny differences in electric current are used to record whether one or more binary digits is registering a value of 1 or 0, and based upon these inputs, various *logic gates* produce corresponding outputs. Traditionally, these are the *AND*, *NOT* and *OR* gates,[4] which many will have encountered in high school physics or in an introduction to electronics.

> ```
> 1010101
> 0101010
> 1010101
> ```
> **Vector spaces over $\mathbb{Z}_2$**
>
> The binary values $\{0, 1\}$ have their own algebraic structure and can be thought of as a number system or *field* in their own right, with addition defined by the formulas
>
> $0 + 0 = 1 + 1 = 0, \quad 1 + 0 = 0 + 1 = 1$
>
> and multiplication by the formulas
>
> $0 \cdot 0 = 0 \cdot 1 = 1 \cdot 0 = 0, \quad 1 \cdot 1 = 1.$
>
> This number field is called $\mathbb{Z}_2$ or sometimes K, which enables 'words' consisting of n binary digits to be regarded as points in the *binary vector space* $\mathbb{Z}_2^n$ or K^n.
>
> The theory of binary vector spaces is one of the backbones of modern information theory, coding theory and cryptography (see Shannon, 1948, Hamming, 1980, Hankerson et al., 2000).

---

[4]In practice it is more efficient to use only *NOR* gates (i.e. *NOT*(a *OR* b)), since all the other logic gates can be built from combinations of *NOR* gates, an interesting exercise.

TABLE 7.1 The three basic logic gates or Boolean connectives: **NOT** (Negation), **AND** (Conjunction) and **OR** (Disjunction)

a	**NOT** a	a	b	a **AND** b	a	b	a **OR** b
1	0	1	1	1	1	1	1
0	1	1	0	0	1	0	1
		0	1	0	0	1	1
		0	0	0	0	0	0
Negation		Conjunction			Disjunction		

Their standard *truth tables* are given in Table 7.1.[5] We can use these connectives to write equations like (7.16) even more explicitly as

$$w = t \textbf{ AND } s \textbf{ AND } (p \textbf{ OR } r).$$

These three logical connectives can be used to combine stand-alone symbols such as a and b into compound statements such as a **AND NOT** (b **OR** c). The combination rules in Table 7.1 can be used to deduce general laws such as the *De Morgan equivalence*

$$(\textbf{NOT } a) \textbf{ AND } (\textbf{NOT } b) = \textbf{NOT } (a \textbf{ OR } b), \qquad (7.17)$$

and from these the science of *Boolean logic* develops. As with Euclidean geometry and linear algebra, Boolean logic can be defined through a set of basic laws (axioms) from which other laws (theorems) can be deduced (see Hamilton, 1982, p. 106).

The usefulness of Boolean algebra in search engines stems from a very important equivalence between the Boolean connectives and the operations of set theory (introduced in Section 1.2). The operation of set *union* $A \cup B$ corresponds to Boolean *disjunction*, the operation of set *intersection* $A \cap B$ corresponds to Boolean *conjunction*, and the set *complement* (written by Boole using the equation $1 - A$, and today often written A') corresponds to Boolean *negation* (see Table 7.2). That every collection of sets can be described by Boolean algebra in this way is relatively easy to see from the way set union, intersection and

---

[5] The terms *negation* for **NOT** and *conjunction* for **AND** are probably familiar, especially since the word *and* is a conjunction in elementary grammar as well as in logic. The term *disjunction* for **OR** is less familiar in non-technical fields and can sound off-puttingly similar to *dysfunction*. Don't be put off: disjunctions are the most inclusive of the connectives, some being more inclusive than others (as shown a few pages ago in Figure 7.1).

TABLE 7.2 The correspondence between the basic operations of set theory and the connectives of Boolean logic. The equations on the right make the relationships explicit, and can be used to define the set-theoretic operations.

Set theory	Boolean logic	Equivalence / Definition
Complement	Negation	$A' \iff$ things *NOT* in A
Intersection	Conjunction	$A \cap B \iff$ things in A *AND* in B
Union	Disjunction	$A \cup B \iff$ things in A *OR* in B

complement are defined. The converse statement is much more striking: every possible configuration of algebraic symbols which obeys the truth tables in Table 7.1 is equivalent to a collection of subsets of some set U.[6]

Consider now a set U of natural language documents. For any keyword a, we can define the subset $A \subseteq U$ of all documents which contain the keyword a. For any list of keywords joined together by a suitable arrangement of *NOT*, *AND* and *OR* connectives (called a *Boolean query*) to a subset of the document collection U using the equivalences in Table 7.2. For example, if s, t, p and r were four keywords, we could find the subset W of the document collection that corresponds precisely to the Boolean query $w = t$ *AND* s *AND* $(p$ *OR* $r)$, corresponding (at least in theory) to Boole's definition of *wealth* in equation (7.16). In a Boolean search engine, this subset W is then returned to the user, these documents being potentially relevant to the query w.

This method has its limitations, as discussed in Chapter 5. Using only the numbers 0 and 1 to measure relevance gives no way of ranking the matching documents according to different degrees of relevance, and the assumption that every document that matches a Boolean query is actually relevant to the user who wrote the query is often mistaken. The method also fails to find documents which don't contain a user's keywords but which are still extremely relevant. A Boolean search engine has no way to match the query *heart attack* to documents containing the term *cardiac arrest* or *myocardial infarction* without a

---

[6]This equivalence was proved by M.F. Stone and is called the *Stone Representation Theorem*. It was not proved until 1933, and requires the Axiom of Choice (Birkhoff, 1967, Ch 12).

detailed thesaurus. Unlike the WORDSPACE models, it's hard to navigate from a near miss to a hit using Boolean logic because there is no notion of nearby or 'almost relevant,' because the relevance scores can only take the values zero and one.

Boolean search engines are often ideal for a user who knows the terminology of a particular field and can form a good query. The simplicity of Boolean logic (and its relative familiarity to many people) makes such a search engine fairly transparent to use, so that Boolean searches can be designed and modified with considerable control and accuracy, at least by an expert user. For these reasons, almost every specialized search engine (such as a library catalogue) supports full Boolean queries, and many internet search engines have *Advanced Search* options allowing you to enter conjunctions (*"All of these words"*), disjunctions (*"Any of these words"*) and negations (*"None of these words"*).

7.2 Vector Negation in WORDSPACE

We've seen that a WORDSPACE model can be used to explore a general area of meaning, and that by contrast Boolean logic specifies a particular subset of documents crisply and efficiently, but inflexibly. Is there any way we could combine the deductive power of logic with the more flexible and inductive knowledge modelled in WORDSPACE? It seems like a tall order. So far our operations using vectors have been pretty primitive: we have added two (or more) word vectors together to get a variety of query vectors and document vectors, and we have used the Euclidean distance or the cosine similarity measure to find which words are closely related to one another. Between them, these two operations allow an information retrieval system to formulate a simple additive query statement and to search for nearby documents, which is impressive in its way but very blunt. One standard semantic objection to this framework is that these vector operations provide no model for negation, which is a gaping omission — as Boole himself wrote:

> Thus, we cannot conceive it possible to collect parts into a whole, and not conceive it also possible to separate parts from a whole.
>
> (Boole, 1854, p. 33)

This section describes a solution to this problem by giving a sound method for negation in WORDSPACE, a step which leads this research into the world of quantum logic. The problem (and an example of its

TABLE 7.3 Removing the musical meaning from *rock*, leaving the geological meaning, in the BNC WORDSPACE

rock		rock NOT pop	
rock	1.000000	rocks	0.728442
pop	0.695325	rock	0.718695
band	0.691613	sand	0.687207
rocks	0.665700	mud	0.603374
jazz	0.655632	limestone	0.595006
music	0.622746	ice	0.553469
singer	0.589253	lava	0.527247
musician	0.560757	sediment	0.524154
guitarist	0.556149	gravel	0.518139
punk	0.549744	boulders	0.517634
indie	0.536820	mountain	0.512865
sand	0.536614	sediments	0.508641
song	0.535376	shallow	0.506559
musical	0.526710	beneath	0.499698
drummer	0.525065	granite	0.499528
ice	0.521718	lakes	0.499374

successful solution) is illustrated in Table 7.3. The left hand column of this table shows the nearest neighbours of the word *rock* in the BNC WORDSPACE, all of which are about *rock music*. When we remove the word *pop* from this query, the right hand column shows that we have obtained the physical meaning of *rock* instead. This removal operation in WORDSPACE will be called *vector negation*.

7.2.1 Orthogonality and Irrelevance

Vector negation depends on another crucial geometric idea, that of *orthogonality*. *Orthogonal* means the same thing as *perpendicular* or *at right angles* — the only difference is that the term orthogonal is used more often when talking about higher dimensions. Fortunately, the mathematics of vectors which we developed in Chapter 5 can be used to translate the geometric idea of two points a and b being perpendicular to one another into very simple algebraic equations about their coordinates $(a_1, a_2, \ldots, a_n)$ and $(b_1, b_2, \ldots, b_n)$.

It two lines are at right angles, then the angle between them is 90 degrees, and the cosine of this angle is zero (Figure 4.4, page 106). It follows that two lines are perpendicular if the cosine similarity of the vectors representing those lines is equal to zero, and thanks to the method of working with vectors in any dimension using their coordinates, this translates easily into a simple equation with numbers that we can run on a computer

Throughout this chapter, we'll be working with normalized vectors, and for these vectors the scalar product $a \cdot b$ and the cosine similarity measure $\cos(a, b)$ are the same thing. We'll write this as $a \cdot b$ (or, for example, *bank* $\cdot$ *bass* for the word vectors of Chapter 5) rather than $\cos(a, b)$, since the notation $a \cdot b$ is easier and more in line with the books on linear algebra to which you might turn for more information.

Definition 16 (Jänich, 1994, §8.2)

Two (non zero) vectors $a = (a_1, a_2, \ldots, a_n)$ and $b = (b_1, b_2, \ldots, b_n)$ are said to be perpendicular or *orthogonal* if their scalar product is equal to zero, i.e.
$$a \cdot b = 0.$$

Example 8 Orthogonality of word and document vectors

To get more of a feel for what orthogonality is all about, recall the word vectors and document vectors of Section 5.6, including (from page 158):
$$\textit{bank} = (0, 0, 1), \quad \textit{bass} = (0.477, 0.894, 0)$$
and (from page 161):
$$\textit{Doc1} = (1, 0, 0), \quad \textit{Doc2} = (0, 1, 0) \quad \text{and} \quad \textit{Doc3} = (0, 0, 1).$$
Since we have
$$\textit{bank} \cdot \textit{Doc1} = (0 \times 1) + (0 \times 0) + (1 \times 0) = 0$$
it follows that *bank* and *Doc1* are orthogonal, according to Definition 16. In the same way, we can deduce that *bank* and *Doc2* are orthogonal, and that *bass* and *Doc3* are orthogonal. It's also easy to see that the two terms *bank* and *bass* are orthogonal, since
$$\textit{bank} \cdot \textit{bass} = (0 \times 0.477) + (0 \times 0.894) + (1 \times 0) = 0.$$

Figure 7.2 shows the relationship between *bank* and *bass* (and between *commercial* and *guitar*) pictorially, from which we can clearly see that 'having a scalar product of zero' and 'being at right angles to one another' are the same property.

LOGIC WITH VECTORS: AMBIGUOUS WORDS AND QUANTUM STATES / 211

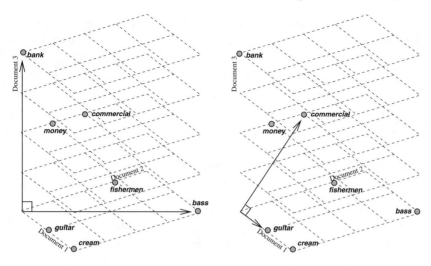

FIGURE 7.2 This diagram shows that the word vectors for *bank* and *bass* are perpendicular to each other, as are those for *commercial* and *guitar*.

Notice that in the three documents example, two terms are orthogonal precisely when they *never occur together* in the same document. This is just what we should expect from the equation $a \cdot b = 0$. Since all our coordinates are positive in this example,[7] the only way the sum $\sum a_i b_i$ can come out to be zero overall is if every term in this summation is zero: so for every i, we must have $a_i b_i = 0$, which implies that at least one of a_i and b_i is equal to zero. It follows that whenever the vector a has a nonzero coordinate in the ith position, b must have a zero coordinate in the ith position, and vice versa. In other words, the vectors a and b must have no features in common at all for them to be orthogonal to one another.

Now suppose that a represents a query vector and b a document vector, and that we're using cosine similarity to measure the extent

---

[7]If coordinates measure purely and simply the number of times a given word occurs in a given document, all these coordinates will be positive whole numbers or zero — a word can't occur a negative or a fractional number of times. However, with some dimension reduction techniques (including singular value decomposition), we sometimes end up with vectors that have both positive *and* negative coordinates. For example, if we added a second latent axis to Figure 6.2 on page 176, it would be used to measure the extent to which each point deviates from the first or principal axis, and since this deviation could be to *either* side of the principal axis, we would need to use both positive and negative coordinates to distinguish *which* side a point is on.

to which the document b is relevant to the query a (as is standard in retrieval systems built on the vector model). If a and b are orthogonal then the system gives the answer $a \cdot b = 0$, telling us that the document b has no coordinates which match those of the query a at all. In this case we should conclude that b is completely *irrelevant* to our query a. This rule for irrelevance can be applied to any vectors in WORDSPACE, and is summed up by the following definition:

Definition 17 Two word vectors a and b in WORDSPACE are considered *irrelevant* to one another if their vectors are orthogonal, i.e. a and b are irrelevant to one another if $a \cdot b = 0$.

Just as with many of the geometric relationships we have encountered in this book, irrelevance is a relationship between a *pair* of objects. A document can't be irrelevant all on its own — it has to be irrelevant *to* something else.

This puts us in a position to properly address the question, "How do we form a vector for the query a *NOT* b in WORDSPACE?" To model the meaning of a statement like *rock NOT pop* in such a way that the system realizes we are interested in the geological meaning, not the musical meaning of the word *rock*, we need to find those aspects of the meaning of *rock* which are different from, and preferably unrelated to, those of *pop*. Meanings are unrelated to one another if they have no features in common at all, just as a document is regarded as completely irrelevant to a user if its scalar product with the user's query is zero — precisely when the query vector and the document vector are orthogonal.

The initial intuition behind vector negation is quite simple — if we add a new word b to a query a by *adding* its vector, then surely we should be able to remove the word b by *subtracting* its vector. But it turns out that subtracting the whole of vector b to give the query vector $a - b$ can have disastrous results, removing many of the parts of a which you wanted to keep and returning documents that are further than ever from the ones you wanted. Instead, we need to rescale b by some real number λ (remember that we can always rescale a vector in this way) to give a new vector $a - \lambda b$. This leaves the question of what the number λ should be.[8] Ideally, we should subtract exactly the right amount to make the

---

[8]This question is also addressed in Salton and Buckley (1990) and Dunlop (1997). For a more thorough discussion and analysis see Widdows (2003b).

unwanted vector b *irrelevant* to the results we obtain, so according to Definition 17 we need to make our final query vector *orthogonal* to b.

In this way, we deduce that

1. The vector for a **NOT** b should take the form $a - \lambda b$ for some number $\lambda \in \mathbb{R}$.
2. The vector a **NOT** b should be orthogonal to b, so that $(a \, \textbf{NOT} \, b) \cdot b = 0$.

This is all we need to know to work out the number λ, which turns out to be given by the following equation (Widdows, 2003b, Theorem 1):

$$\lambda = \frac{a \cdot b}{||b||^2}.$$

This is particularly easy to see in the case of normalized vectors, since in this case the norm $||b||$ is equal to 1 and we get the result that

$$a \, \textbf{NOT} \, b = a - (a \cdot b)b. \qquad (7.18)$$

Taking the scalar product of this expression with the vector b, we find that

$$\begin{aligned}(a \, \textbf{NOT} \, b) \cdot b &= (a - (a \cdot b)b) \cdot b \\ &= a \cdot b - (a \cdot b)(b \cdot b),\end{aligned}$$

and since $b \cdot b = 1$ for normalized vectors, this reduces to the expression $a \cdot b - a \cdot b$ which cancels out to zero: exactly what we need for b to be irrelevant to a **NOT** b. In practice the vector a **NOT** b is then renormalized (that is, scaled back up to its original size), so that we can fairly compare similarities for the new vector a **NOT** b with those of any other vector, just as in Section 5.5.4.

Example 9 *bass NOT fishermen*

As a concrete example, consider the plight of a user trying to find documents about *bass players* in our familiar three document retrieval system. Recall from Table 5.2 that the word vector for *bass* is $(0.447, 0.894, 0)$. If this query was used to search for documents, we would start by retrieving document 2, with its vector $Doc2 = (1, 0, 0)$.

This document is not what our user is looking for at all — it's all about *bass fishing* and certainly not *bass guitars*. This is because the word *bass* is ambiguous: the system is returning documents that are relevant to *bass* as in fish, not to *bass* as in music.

In order to get rid of this unwanted meaning, suppose our user, seeing which results are offered, clicks on an option which *removes* the term *fishermen* = $(0, 1, 0)$. The number λ, given by the scalar product of *bass* and *fishermen*, is equal to 0.894, giving the query vector

$$\begin{aligned} \textit{bass NOT fishermen} &= \textit{bass} - (0.894)\, \textit{fishermen} \\ &= (0.447, 0.894, 0) - (0, 0.894, 0) \\ &= (0.447, 0, 0) \end{aligned}$$

which is then normalized to give the query *bass NOT fishermen* = $(1, 0, 0)$.

Now we're in much better shape — this query vector will be matched to Document 1 which *is* about *bass guitars*.

7.2.2 Examples and experiments with vector negation

In this section we give some live examples of vector negation in WORDSPACE, and describe the results of experiments comparing vector negation with Boolean negation.

If we look up the neighbours of the word *suit* in a WORDSPACE built from NYT data, we'll find that they are all about the legal meaning of the *suit* (see the left hand column of Table 7.4). Suppose instead that you wanted to find a store where you could buy or clean your *suit*. Then you'd want to be able to *remove* the predominant legal meaning and be left with documents related only to the clothing meaning of *suit* (just as in the right hand column of Table 7.4). You can do precisely this by clicking on the *Subtract from query* box for *lawsuit* in the Infomap demo.[9] Note that you don't have to list unwanted terms one by one — if you just remove the term *lawsuit*, all of the other legal terms disappear as well.

An experiment exploring these promising properties strengthened the hypothesis that vector negation removes whole areas of meaning. The experiment involved manufacturing 1200 queries of the form

keyword a *NOT* keyword b,

and comparing the documents retrieved from WORDSPACE using vector negation with those retrieved using the Boolean method (finding

---

[9] infomap.stanford.edu/webdemo

TABLE 7.4 Neighbours of the vectors for *suit* and *suit NOT lawsuit* in a WORDSPACE built from NYT data

suit		suit *NOT* lawsuit	
suit	1.000000	pants	0.810573
lawsuit	0.868791	shirt	0.807780
suits	0.807798	jacket	0.795674
plaintiff	0.717156	silk	0.781623
sued	0.706158	dress	0.778841
plaintiffs	0.697506	trousers	0.771312
suing	0.674661	sweater	0.765677
lawsuits	0.664649	wearing	0.764283
damages	0.660513	satin	0.761530
filed	0.655072	plaid	0.755880
behalf	0.650374	lace	0.755510
appeal	0.608732	worn	0.755260
action	0.605764	velvet	0.754183
court	0.604445	wool	0.741714
infringement	0.598975	hair	0.739573
trademark	0.592759	wore	0.737148
litigation	0.591104	wears	0.733652
settlement	0.585363	skirt	0.730711
alleges	0.585051	dressed	0.726150
alleging	0.568169	gown	0.723920

documents with keyword a and then removing any document from these results that contains the unwanted keyword b). This revealed some interesting differences between the two: while the Boolean method is (by definition) the most reliable method for making sure you never set eyes on the keyword b in the retrieved documents, vector negation also performs this task very well. Then I took the task a step further, measuring the performance of each method at removing the WordNet synonyms and WORDSPACE neighbours of the unwanted keyword, reasoning that if a user explicitly removes a term, the last thing they want is to start finding synonyms for the same term as an alternative. Vector negation was the clear winner at these tasks, returning 20% fewer WordNet synonyms and 74% fewer WORDSPACE neighbours of the unwanted keyword compared with the Boolean removal method (this difference became even more noticeable for queries where more than one unwanted keyword was removed). These results are described in greater depth in Widdows (2003b).

The case of *suit NOT lawsuit* in the New York Times (Table 7.4) is a good example. All of the top documents returned for the query *suit* are about legal actions, and even the ones that don't actually contain the word *lawsuit* (which would be retrieved by a Boolean search for *suit NOT lawsuit*) are *about* lawsuits. On the other hand, the top documents for the *vector* version of *suit NOT lawsuit* are all about fancy clothing — even those that don't actually contain the word *suit*. This demonstrates that while the Boolean method has more 'local control' over search outcomes, the vector negation method has a more widespread influence over whole regions of WORDSPACE.

7.3 Ambiguity and the Quantum

Why might we expect a projection operator on WORDSPACE to have anything in common with quantum mechanics? To understand the link between ambiguity and quantum theory, we will start with Aristotle's description in the 'philosophical lexicon.'

> 'Quantum' means that which is divisible into two or more constituent parts, each of which is by nature a 'one' and a 'this'.
> (*Metaphysics*, Bk. V, Ch. 13)

Interestingly, this definition is followed immediately by a rare passage where Aristotle discusses the issue of measurement:

> A quantum is a plurality if it is numerable, a magnitude if it is a measurable. Plurality means that which is divisible potentially into non-continuous parts, magnitude that which is divisible into continuous parts;

As we shall soon see, measurement has become a central (and strange) issue in modern quantum mechanics, and it is curious (if possibly accidental) that Aristotle describes the quantum and measurement in almost the same breath.

Though Aristotle's description predates the founding of modern quantum theory by two and a quarter millenia, there is an important common thread. Quantum theory involves dealing with particles which are composed from different pure states, which can be superimposed upon one another to make combined states. In the same way, ambiguous words can be thought of as the sum of different 'pure' meanings, superimposed upon the same word. In modern quantum theory, the quanta are usually the units into which the composite particle is broken down, whereas for Aristotle, the combination itself is the quantum object — otherwise, the frameworks have much in common.

In the quantum view, elementary particles such as electrons and photons (light particles) are represented by basic wave patterns which are essentially three-dimensional versions of the harmonic waves in Figure 1.4. Given a suitable scalar product, these basic harmonic waves are actually *orthogonal* to one another, and so they make good basis vectors for a vector space, and each particle can be represented by a set of coordinates which measures the extent to which each resonant frequency (or energy) contributes to the whole particle.

Because these basic building blocks are waves, the representation of a particle in quantum mechanics becomes spread out over a whole region of space, and in quantum mechanics particles can therefore interact over great distances. By contrast, interactions in classical physics are limited to a local area which is firmly fixed by the distance a light beam can travel in a given time. Is there any sense in which our Boolean negation operator, which removes precisely those documents which actually *contain* a particular word, is analogous to an interaction in classical physics? On the other hand, is there any sense in which our vector negation operator, which removes documents which are anywhere in a whole negated *region* of WORDSPACE, is analogous to an interaction in quantum physics?

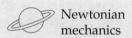 Newtonian mechanics

Newton's laws of motion and theory of gravitation are cornerstones of classical physics. Newton's laws recognize that force is proportional to *acceleration* (in contrast to Aristotle's view that force is needed to cause motion itself). Knowledge of the forces acting on an object therefore tells us how the object will accelerate.

Because acceleration measures change in velocity and velocity measures change in position, we need to know both the initial velocity and initial position of the object to predict its future behaviour exactly. However, according to quantum mechanics, it is impossible to measure both position and velocity (or momentum) simultaneously and obtain exact results, so it is impossible to predict the object's future exactly.

The analogy between quantum particles and ambiguous words turns out to be quite strong — as well as being appealing on a general intuitive level, the exact same operations in vector spaces can be used to model both processes (though due to some remarkable matrix symmetries, we have reason to expect these operations to be more accurate models in physics than they are in linguistics). To take the reader on this part of the journey, it will be fruitful first to introduce some of the concepts behind modern quantum mechanics.

7.3.1 Some concepts from quantum mechanics

Quantum mechanics and the theory of relativity are often described as the two great discoveries of physics in the twentieth century, and of these, quantum mechanics poses by far the biggest challenge to the conceptual system which we call classical physics. The theory of special relativity can be obtained by adopting the *Minkowski metric* (page 110) which intertwines space and time coordinates into one measure of distance in *spacetime*, and the theory of general relativity can be obtained by allowing this spacetime to be *curved* rather than flat. While the mathematics behind this process is difficult, classical physics adapts very elegantly to its new setting. Newton's laws of motion are shown to be correct approximations to the predictions of general relativity, and are still valid unless you are near to a very heavy object or travelling very quickly indeed.

Quantum mechanics, on the other hand, flies full in the face of classical physics. At its core, classical physics gives a *deterministic* model for the universe: if the state of the universe at time t is known, and the laws that govern the change or *evolution* of this state are also known, then is is possible to predict the state of the universe at any future time t_+. It is also possible to reconstruct the state of the universe at any past time t_-, so classical mechanics can be used (for example) both to predict future astronomical phenomena such as the next appearance of Halley's comet, and to corroborate records of its past appearances throughout history.

This deterministic certainty is absent from quantum mechanics. Quantum physics does not predict the exact state of a particle: instead, it gives the *probability* that the particle will be found to be in a particular state if it is observed. This uncertainty or non-determinism still poses a great challenge to physics and philosophy, as summed up by another of Einstein's famous quotes,[10]

---

[10]It should be emphasized that Einstein was not opposed to quantum mechanics, and was instrumental in its development (Wheeler, 1991) — he just found it to be very bizarre, which it is.

> I shall never believe that God plays dice with the world.

In spite of Einstein's misgivings, many experiments have demonstrated that God *does* play dice with the world, and that the future (and indeed the present) is composed of a variety of different possible states or outcomes.

For example, in the double-slit experiment (opposite), it is impossible to predict with certainty whether the particle will pass through the slit A or the slit B. Not only this: it's also impossible to account for the particle's behaviour without assuming that it went through *both* slits at once! More precisely, the particle is represented by a sum of different *waves*, some of which go through A and some of which go through B. (It's much easier to imagine a wave passing through two slits simultaneously than a point particle doing the same thing.)

Strangely enough, if we place a detector across each of the slits and use this to observe which path the particle takes, then the *complete* particle will be observed to pass through one or other of the two slits, not both — an observation of this sort never yields the mixed state of affairs (part A and part B)

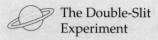

The Double-Slit Experiment

One of the most famous successes of quantum theory is its explanation of the following experiment. A single particle is fired from the emitter E towards a barrier with two slits A and B. If the particle passes the barrier, it continues on to reach the screen S, where its position is observed.

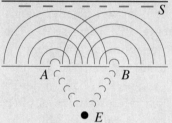

The distribution of particles reaching the screen S has distinctive peaks (shown by the grey bars) called *interference fringes*. These can be explained with astonishing accuracy by quantum mechanics, which models the particle as a *wave*. Part of this wave passes through A, part of it passes through B, and these two parts interfere with *each other*, giving the interference pattern on the screen S. In other words, the particle has passed through *both* slits at once, which is impossible in classical physics because the particle is represented by a single point not a dispersed wave.

that accounts for the interference fringes. In such a situation where we know for sure which slit the particle passed through, the interference patterns are not observed: it is almost as if the particle knows that it's being watched and so behaves itself.

This is curiously similar to the abilities of humans to distinguish between the meanings of words — which seems so effortless and obvious to us. Even if a word could potentially have many meanings, the moment we encounter it in context, we usually know exactly which meaning is intended. The interpretation of meaning is often so natural for us that the suggestion of different possible interpretations is often an opportunity for humour, rather than the result of confusion. Other overt and apparent similarities between observations in quantum theory and word-sense disambiguation are described by Gabora and Aerts (2002) and Widdows (2003a).

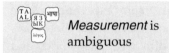
Measurement is ambiguous

Measurement, and many similar words such as *diagnosis, discovery, invention,* and even *writing* and *construction,* can be used to refer to both an activity and the outcome or result of that activity. This is a typical case of systematic ambiguity, whereby whole classes of words can take a variety of meanings, where these meanings are in a sense different sides of the same coin (Pustejovsky, 1995).

In classical physics, it was often thought that measurements could be objective, so that the *outcome* of a measurement was independent of the *activity* of measurement: these two aspects of the meaning of *measurement* could be separated. Thanks to quantum mechanics, we know that this is not the case.

7.3.2 Vectors for quantum mechanical states

How are we to model a quantum particle that behaves as if it has been in two places at once? The mathematical solution to this problem was developed by Paul Dirac (1902–1984), who presented a formal axiomatic approach to quantum mechanics in his famous work, *Principles of Quantum Mechanics* (1930). In this approach, the state of any system is represented by a *state vector*. For example, if a particle in the double-slit experiment (shown in the box on page 219) is known to have passed through the slit A, this would be represented by the symbol $|A\rangle$, and if it passed through the slit B, this would be written by the symbol $|B\rangle$. Each of these possibilities is called a *pure state*. But these are not the only possibilities: there are also any number of hybrid possibilities where part of the particle (or wave) goes through A and part goes through B, and these two parts then interfere with one another. The state of such a hybrid particle is represented by the combination

$$\alpha_1|A\rangle + \alpha_2|B\rangle, \qquad (7.19)$$

where α_1 and α_2 are numbers.[11] If we think of the pure states $|A\rangle$ and $|B\rangle$ as basis vectors, and the numbers α_1 and α_2 as coordinates, then the combined state represented in equation (7.19) is also a vector. We assume that the pure states $|A\rangle$ and $|B\rangle$ each have unit length, and that $|A\rangle$ and $|B\rangle$ are orthogonal to one another, which in Dirac's notation is written

$$\langle A|A\rangle = \langle B|B\rangle = 1, \langle A|B\rangle = 0.$$

Dirac's notation $\langle A|B\rangle$ for the scalar product is similar to the notation $[A|B]$ used by Grassmann for his 'inner products,' nearly 100 years previously (Grassmann, 1862, §137), from which the scalar product can be derived. Just as in WORDSPACE, the state vectors are normalized, so that

$$\alpha_1^2 + \alpha_2^2 = 1.$$

Quantum computing

Equation (7.19) is used to represent the state of a *quantum bit* or *qubit* in the infant field of quantum computing. In theory, many important algorithms could run much more efficiently using qubits than is possible using traditional binary bits: for example, quantum computing could revolutionize search problems (including mathematical problems such as finding the prime factors of a large integer) by enabling several possibilities to be explored simultaneously (Hirvensalo, 2001).

---

[11] In quantum mechanics, the coordinates α_j are *complex* numbers, though the most significant property of these coordinates is their *magnitude* or length, which is a real number. The adaptation of vector spaces to use complex numbers as coordinates is standard, and complex numbers make things easier because they enable more equations to be uniquely solved. For the sake of simplicity we will stick to real numbers in this book; for a proper introduction to quantum mechanics and measurement in complex Hilbert spaces, see Cohen (1989).

222 / GEOMETRY AND MEANING

This vector approach allows multiples of the wave functions for the states $|A\rangle$ and $|B\rangle$ to be *superimposed* on one another, just as the harmonic waves in Figure 1.4 can be superimposed on one another to yield a new wave. In this way, the vector approach enables us to model the interference patterns seen in the double-slit experiment.

As explained in Section 7.3.1, if the state of the particle is observed by placing detectors over the slits A and B, only the pure states $|A\rangle$ and $|B\rangle$ will be encountered. Somehow, the combined quantum state $\alpha_1|A\rangle + \alpha_2|B\rangle$ becomes 'quantized' into one of the pure states $|A\rangle$ or $|B\rangle$. The probability of each of these outcomes is given by the numbers α_1^2 and α_2^2. This explains why quantum mechanics uses *normalized* vectors: it ensures that for any vector, the sum of the squares of the coordinates ($\sum \alpha_i^2$) adds up to 1, which is necessary if the magnitudes α_i^2 are to represent probabilities.[12]

7.3.3 Vector negation in terms of quantum states

The model for vector negation in WORDSPACE, developed independently in Section 7.2, can be derived from similar hypotheses. Suppose that the word vector for an ambiguous word w receives contributions from just two pure senses w_1 and w_2, so that our ambiguous word vector w can now be thought of as a combination of the form

$$w = \alpha_1 w_1 + \alpha_2 w_2, \qquad (7.20)$$

where w, w_1 and w_2 are unit vectors and $\alpha_1^2 + \alpha_2^2 = 1$.

Suppose that a user is only interested in the sense w_2, and so constructs the query w **NOT** w_1. This gives us values for the vectors w and w_1 in equation (7.20), which can be solved to find w_2 by using the substitution $\alpha_1 = w \cdot w_1$.[13] It follows that

$$\begin{aligned} w_2 &= w - \alpha_1 w_1 \\ &= w - (w \cdot w_1) w_1, \end{aligned}$$

---

[12]More information on the interpretation of quantum mechanics as a calculus of probabilities is given by Wilce (2003), whose article is available online through the Stanford Encyclopedia of Philosophy.

[13]This is a standard projection method for finding the coordinate of w along the w_1-axis (Jänich, 1994, p. 31). In this way, the coordinate α_1 is easily calculated using cosine similarity. In basic coordinate geometry, α_1 is sometimes described as the *direction cosine* of w in the w_1 direction, and it can be calculated from w and w_1 using the cosine similarity measure.

which, once we have exchanged the symbols w and w_1 for a and b, is exactly the same as equation (7.18) on page 213.

In this way, the same predictions are obtained in both quantum mechanics and WORDSPACE, because they are derived using the same models. In assessing their value, we must question the assumptions upon which these models rest. For example, there are good reasons in physics for assuming that the pure states w_1 and w_2 are orthogonal to one another — this crystalline perpendicular structure is already built into the theory through the symmetries of the harmonic waves used to model pure states. In natural language, things are much more difficult — the idea of two different meanings being *totally* irrelevant or orthogonal to one another seems unlikely, if indeed it ever happens at all. It is not difficult to find familiar sense-distinctions between which vector negation in WORDSPACE does not discriminate very well, partly because so much ambiguity is systematic, and the orthogonality assumption used to solve equation (7.20) is too strong. One alternative used by Schütze (1998) was to use sense-vectors to represent the 'pure states' of the system, these sense-vectors being the centroids of clusters of sentence-vectors derived empirically from the original corpus.

On the whole, the WORDSPACE software does a surprisingly good job of finding interesting ambiguities and tracking down regions corresponding to different senses, and the statistical behaviour of the vector operators in larger experiments supports this conclusion. The behaviour and potential of vector operators in WORDSPACE is investigated more thoroughly by Widdows and Peters (2003) — the structure becomes increasingly interesting as we try to add the other logical connectives to the system.

7.4 From points to lines and planes

The previous section described how a vector representing an ambiguous word in WORDSPACE can be divided into its component parts. Conversely then, how can points representing different parts of a concept be assembled into a whole? Yet again, we will see that Aristotle points the way, and the geometric tools developed by people like Euclid, Descartes and Grassmann enable us to turn Aristotle's suggestions into working geometry and algebra.

The idea of combining points in WORDSPACE that represent specific parts into any *single* point representing a more general whole has an essential flaw. By adding together their Cartesian coordinates, we can combine two points a and b in a vector space to give their vector sum $a + b$ (page 134), but while this works well for modelling some query operations, there is no sense in which the point $a + b$ *contains* either the point a or the point b (unless a and b are identical). Just as in Euclid's original definition of points (page 9), $a + b$ can be described as a point precisely because it *doesn't* have any parts.

One of the drawbacks of vector space models which consider only points is that no point is more or less general than any other — if no point can contain any other, then no point can represent a concept which contains another concept. This is a serious drawback, because such a model can not express the way in which the term *elephant* is more specific than the term *mammal* which is itself more specific than the general term *animal*. If we consider only points, all the efforts in Chapter 3 to understand and model hierarchical relationships become lost in WORDSPACE.

As well as being an issue when trying to represent information taxonomically, this problem is also central to logic, because of the correspondence between geometric containment and logical implication. For example, consider the challenge of finding a geometric representation for the logical disjunction *OR*. In set theory and the corresponding Boolean logic, the disjunction of two points is modelled by their set union (see Table 7.2), that is to say, the points a and b are thought of as singleton sets $\{a\}$ and $\{b\}$, and their disjunction a *OR* b corresponds to the set union $\{a\} \cup \{b\} = \{a, b\}$. In this way, a *OR* b manages to contain both a and b, because a *OR* b is no longer just a single point — it is a set.

In a similar way, more general concepts which contain their more specific ingredients can be constructed in other mathematical models, including vector models like WORDSPACE, and (in low dimensions) the shapes we use are already familiar concepts — lines, surfaces, and solids. Following Aristotle's discussion of the quantum just a little further, after he describes quanta as either discrete pluralities or continuous magnitudes, we read that

> of magnitude, that which is continuous in one dimension is length; in

two breadth, in three depth. Of these, limited plurality is number, limited length is a line, breadth a surface, depth a solid. (*Metaphysics*, Bk V, Ch. 13)

Since a point can be contained in a line, a line contained in a surface, and a surface contained in a solid, this gives us a way of building a logical structure where more specific concepts are represented by regions with fewer dimensions, and more general concepts are represented by regions with more dimensions.

7.4.1 One-dimensional quanta

How can we represent Aristotle's 'quanta' — things which are divisible into constituent parts — from the individual parts of which they are constituted? In geometry, if these parts are two points, then we can draw a line between them, which Euclid's first axiom of geometry (page 8) asserts for straight lines. More generally, joining two points with some shortest path is a characteristic operation in many geometries, where the shortest path between two points is determined using one of a variety of possible distance metrics.

The representation $w = \alpha_1 w_1 + \alpha_2 w_2$ (equation 7.20, page 222) used for both an ambiguous word in WORDSPACE and a state vector in quantum mechanics, is

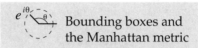 Bounding boxes and the Manhattan metric

When using the Manhattan metric (page 102), the available paths or 'streets' run only in north–south or east–west directions — going diagonally 'as the crow flies' is not allowed. In getting from point A to point B, a variety of shortest paths can be take by making different southward or eastward choices at different parts of the journey.

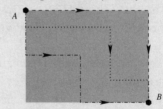

The shaded region between A and B gives the set of all points which lie upon one of these shortest paths, and is the *bounding box* of A and B.

an algebraic version of this joining idea. The points w_1 and w_2 can be thought of as basis vectors, in relation to which the combined vector w is determined by the coordinates α_1 and α_2. If we use normalized vectors (which is standard in both WORDSPACE and quantum mechanics), the coordinates α_1 and α_2 are rescaled so that they satisfy the condition

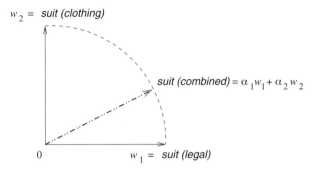

FIGURE 7.3 The ambiguous word *suit* could appear anywhere on the path joining its constituent 'pure' meanings, depending on their relative frequency of use.

$\alpha_1^2 + \alpha_2^2 = 1$. Geometrically, this constrains the vector w to lie on a circle of unit length in the plane spanned by w_1 and w_2.

In WORDSPACE, the word vector w may represent an ambiguous word (such as *suit*), the constituents w_1 and w_2 representing its 'pure' senses (such as the *clothing* and *legal* senses of suit). The coordinates α_1 and α_2 are determined by the number of times the word in question is used with each of its possible meanings in the corpus from which the WORDSPACE is built. Since α_1 and α_2 arise from counting the distributions of actual words, they should not in practice be negative, in which case w is guaranteed to lie on the positive part of the circle lying *between* w_1 and w_2.

The correspondence between the geometric and algebraic ways of representing w are outlined in Figure 7.3. The point where the vector for *suit* actually lies upon this curve was estimated from the results in Table 7.4, which suggest that the legal meaning of *suit* is much more prevalent than the clothing meaning, at least in the New York Times.

The joining of two points into a line enables us to build Aristotle's 'one-dimensional quanta' out of pure states. The Infomap demonstrations and experiments show that such quanta arise naturally from language data, and can sometimes be split into their component parts using spatio-logical operators such as vector negation. The method of using algebraic equations to represent this geometric process is due largely to the work of Descartes. Many intuitive geometric options of this kind have simple algebraic counterparts which are easy to turn into computer programs — for example, if we used the algebraic condition $\alpha_1 + \alpha_2 = 1$ instead of $\alpha_1^2 + \alpha_2^2 = 1$, the point w would be constrained to lie not on a circle but on the straight line joining w_1 and w_2. However, since we have been working with normalized vectors throughout, the circle joining w_1 and w_2 should itself be thought of as a shortest path, just as the shortest flight-path between two points on the earth's surface is given by a great circle on the earth's surface, not by a straight line going through the earth's core. Because of this, it is reasonable to regard the circle in Figure 7.3 as a 'line' in Aristotle's quantum sense, even though it is not a straight line in a Euclidean sense. (In just the same way, it is reasonable to talk about 'lines' on the London Underground, even the 'Circle Line,' which in the relevant context may well be the shortest distance between two points.)

7.4.2 From lines to planes — removing multiple keywords

By Aristotle's time, the generalization from 1-dimensional lines to 2-dimensional planes and surfaces was already well understood. In Book I of the *Elements of Geometry*, Euclid introduces some of the basic concepts as follows.

Definition 5.   A surface is that which has length and breadth only.

Definition 6.   The edges of a surface are lines.

Definition 7.   A plane surface is a surface which lies evenly with the straight lines on itself.

From this we learn that planes have many constituent lines, and that surfaces are bounded by lines (just as lines are bounded by points). The idea of angles between lines and planes unfolds further in Book XI, with the following:

Definition 3.   A straight line is at right angles to a plane when it makes right angles with all the straight lines which meet it and are in the plane.

Some of these concepts are outlined on the left hand side of Figure 7.4. A portion of a plane (shaded) contains the straight lines $0A$ and $0B$, and infinitely many more lines which are parallel or lie evenly in between these two. The line $0C$ is at right angles to *all* of the lines in the plane $0AB$ — in fact, any line perpendicular to two non-parallel lines in a plane will

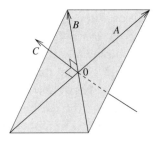 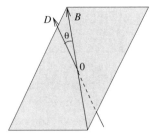

FIGURE 7.4 Planes, some of their constituent lines, and lines orthogonal and inclined to them.

be orthogonal to every other line in the plane as well. Note that we are already talking about planes as composite objects — they can be defined (for example) by three points (such as $0AB$) or by two intersecting lines (such as $0A$ and $0B$). Even the way we write down such concepts implies that lines and planes can be built up by combining smaller units.

Euclid next defines the *inclination* between a line and a plane, describing the angle between that line and its image when projected onto the plane. For example, on the right hand side of Figure 7.4, the line $0D$ projects onto the line $0B$, and the angle θ in between these two lines is the inclination of the straight line $0D$ to the plane. We can use this angle to measure the similarity between lines and planes, and between pairs of planes — in other words, Euclid's *Elements* tells us exactly how to adapt angles and their cosines to measure similarity between higher-dimensional concepts such as lines and planes.

In WORDSPACE, this is useful when you want to remove more than one unwanted keyword from a search. For example, consider the neighbours of *mouse* in Table 7.5 from the point of view of a user trying to find out about its animal meaning. The first set of results is mainly about biological experiments, and since this isn't what you're interested in you can remove the term *cdna* (the name of a specific gene), hoping to avoid this general area. Instead of animals, the neighbours of *mouse NOT cdna* turn out to be related to a *computer mouse* — a different area of meaning altogether! Undaunted, if you also click to remove *printer* from the query, and search again, you find a collection of neighbours related to precisely the sort of *mouse* you were looking for.[14]

---

[14]Unfortunately, *rabbit* and *mice* have been taken out in the cross-fire, an acknowledged danger with negation operators (Kowalski, 1997, p. 160).

TABLE 7.5 Neighbours of the vectors for *mouse, mouse NOT cdna,* and *mouse NOT cdna, printer* in a WORDSPACE built from BNC data

mouse		mouse NOT cdna		mouse NOT cdna, printer	
mouse	1.000000	mouse	0.790258	mouse	0.661624
clone	0.620516	printer	0.541478	dog	0.468058
mice	0.618213	program	0.484000	animals	0.450782
cdna	0.612774	graphics	0.478420	scent	0.422442
gene	0.605464	byte	0.447065	monkey	0.399730
rat	0.601639	colorado	0.436900	humans	0.399563
rabbit	0.592044	apple	0.432437	wild	0.398797
peptide	0.573759	serial	0.428934	animal	0.393381
rna	0.547441	computer	0.428902	dogs	0.393019
myc	0.540731	machine	0.422340	pets	0.390927
assays	0.540298	keyboard	0.417016	goat	0.386561
kb	0.538962	tool	0.408907	rats	0.383797
incubated	0.538298	instruction	0.405817	colorado	0.381239
nf	0.535685	processor	0.405165	mite	0.377886
sequences	0.516982	programmed	0.405074	cat	0.370011
monoclonal	0.513830	notebook	0.403489	rat	0.363774

Mathematically, this is accomplished by making the *mouse* vector orthogonal to the vectors for both *cdna* and *printer*, and thus to the whole plane spanned (i.e. generated) by these vectors. Ideally, this might be accomplished by using the orthogonal projection equation (7.18, page 213), since this equation works successfully for one unwanted keyword. Can we just use this equation twice in succession, once to remove *cdna* and once to remove *printer*? The answer is no, because in practice, these two operations usually interfere with one another — unless the words *cdna* and *printer* happen to be orthogonal to one another (which is very unlikely in practice), the vector negation operations for *cdna* and for *printer* will not commute with one another.

This situation arises in just the same way in quantum mechanics. Two measurements can only be made at the same time if their corresponding projection operators commute with one another, in which case they are said to be *simultaneously measurable* (Polkinghorne, 2002, p. 31). Heisenberg's famous Uncertainty Principle follows because many important operators (such as those for measuring position and momentum) are not orthogonal to one another and their outcomes are entangled.

In WORDSPACE, this problem can be solved by modifying the unwanted vectors w_1 and w_2 to give a pair of vectors $\tilde{w}_1$ and $\tilde{w}_2$ that *are* orthogonal to one another (in the first instance, by setting $\tilde{w}_1 = w_1$ and $\tilde{w}_2 = w_2 - (w_1 \cdot w_2)w_1$, just as in equation (7.18). Any vector orthogonal to the original pair w_1 and w_2 is also orthogonal to the modified pair $\tilde{w}_1$ and $\tilde{w}_2$, because the plane generated by the modified vectors $\tilde{w}_1$ and $\tilde{w}_2$ is the same as the plane generated by the original vectors w_1 and w_2. Thus, by following Euclid's definition and using some algebraic manipulation to get the projections right, we can construct a query vector for *mouse NOT cdna, printer* which is irrelevant to both *cdna* and *printer*. This method for making a set of vectors orthogonal to one another can be generalized to higher dimensions, and it is called the Gram–Schmidt process (Jänich, 1994, p. 142).

The effectiveness of this operation for document retrieval was measured using similar experiments to those described earlier in Section 7.2.2, but now the queries were of the form

$$\text{keyword } a \text{ \textbf{\textit{NOT}} keywords } b_1, b_2.$$

Positive scores for retrieving documents containing the keyword, and negative scores for retrieving documents containing the unwanted keyword and its neighbours and synonyms, were again counted and compared for different methods. In these experiments the differences between vector negation and the Boolean removal method became even more pronounced, vector negation returning 38% fewer unwanted WordNet synonyms and 76% fewer unwanted WORDSPACE neighbours of the negated keyword than the Boolean method.

Based upon this success, a paper (Widdows, 2003b) that implemented a few lines from Books I and XI of Euclid's *Elements of Geometry* was published as new research in a prestigious international conference some 2300 years later. Though fortunate for the author's career, this could only happen in an academic climate that accepts fragmentation as normal. There is hope — the ideas of Aristotle, Euclid, and many other great thinkers are more available today than ever before, and these authors will continue to bequeath many great models and insights to new fields and those as yet uninvented: scholars who take the time to enjoy these ancient and primary sources will always reap rich rewards.

7.5 Subspaces of higher dimension

The adaptation of these Euclidean techniques to higher dimensions, and the geometric algebra which makes this process so general and effective, is inextricably interwoven with the work of Hermann Grassmann (1809–1877). Grassmann, like Boole, was a schoolteacher, but whereas Boole's contributions were recognized and rewarded, those of Grassmann remained obscure and neglected for many years. Many of the ideas of geometric algebra, quantum logic, lattice theory, and much more can be traced explicitly to Grassmann's work. When he first published his theory of many-dimensional quantities (*Ausdehnungslehre*) in 1844, the presentation was too philosophical for the analytic style of mathematics prevalent at the time, and in the 1862 edition Grassmann unwillingly but successfully adapted the work to this 'Definition–Proposition–Proof' format. His work became gradually part of the mainstream, especially when the English mathematician William Kingdon Clifford in 1878 combined the breakthroughs of both Hamilton and Grassmann into a single geometric algebra whose potential and generality is only now being fully realized (Lasenby et al., 2000).

An English version of Grassmann's 1862 *Ausdehnungslehre*[15] is available in most University libraries, and I shall refer to this volume as Grassmann (1862). In this work, Grassmann begins by saying that an *extensive magnitude* α is *numerically derived* from a set of units $\{e_1, \ldots, e_n\}$ if it can be expressed in the form $\alpha_1 e_1 + \ldots + \alpha_n e_n$, the coefficients α_i being real numbers. In the language of vectors developed in Chapter 5, the set $\{e_1, \ldots, e_n\}$ is a basis, the coefficients α_i are coordinates, and this structure is a vector space V. The number n of basic units was defined by Grassmann to be the *order* or *dimension* of the quantity, a pioneering example of the modern definition (Definition 11, page 142).

For example, any point in the plane can be expressed in the form $\alpha_1 e_1 + \alpha_2 e_2$, where e_1 and e_2 represent unit vectors in the x (east) and y (north) directions. Grassmann (1862, §14) calls such a plane the *domain derivable from* the set $\{e_1, e_2\}$. We can choose any two basis vectors e_i and e_j from the set $\{e_1, \ldots, e_n\}$ to construct such a plane, so the total n-dimensional space V contains many different planes of this form (presuming that $n > 2$). Quantities in such a plane can be added together and extended by scalar multiplication without leaving the plane, so such a plane is a vector space in its own right, embedded in the total space V. Effectively, all we have to do is ignore all except two of the coordinates,

---

[15] Translated by Lloyd C. Kannenberg for the American and London Mathematical Society.

and these two coordinates give positions in a plane. Clearly, this process could be applied to any number of coordinates, not just two. Such an embedded vector space is called a *subspace*.

Any vector space of dimension n contains many subspaces of dimension smaller than or equal to n which are vector spaces in their own right. Just as a 1-dimensional line can be embedded in a 2-dimensional plane and a 2-dimensional plane can be embedded in a 3-dimensional space, so any k-dimensional vector space can be embedded as a subspace of an n-dimensional space, so long as $k \leq n$. Only very special subsets of a vector space V are actually subspaces — for a subset $U \subset V$ to be a subspace, the sum of any two points in U must also be in U (so $a, b \in U \Rightarrow a + b \in U$), as must any linear multiple (so $a \in U \Rightarrow \lambda a \in U$).

Thus the set of units $\{e_i, e_j, e_k\}$ is *not* a subspace, but the set of all possible linear combinations $\{\lambda_i e_i + \lambda_j e_j + \lambda_k e_k\}$ which can be derived numerically from $\{e_i, e_j, e_k\}$ *is* a subspace, which is called the *linear span* or just the *span* of the set $\{e_i, e_j, e_k\}$. Thus the linear span can be used to build a subspace $\overline{U}$ from any set of points in $U \subset V$. For example, the plane $0AB$ (see Figure 7.4) is the linear span of the lines $0A$ and $0B$. The linear span is a way of implementing Euclid's requirement that all the parts of a line or plane should lie evenly with themselves.

7.5.1 Projective geometry and algebra

Every subspace of U of a vector space V must pass through the origin point 0. To see this, recall that for any vector $u \in U$, λu must also be in U for every real number λ. Since 0 is a real number and $0u = 0$, the origin point 0 must be contained in the subspace U.

There is an immediate practical benefit of this — we no longer need to use the prefix 0 when describing lines and planes because it is redundant. This goes back to the point in Section 5.2 that a vector can be regarded as either a journey (a beginning and a destination point) or as a point (the destination point of a journey whose starting point, the origin, is predetermined). For example, the line $0C$ in Figure 7.4 can just be called "the line C," and the plane $0AB$ can just be called "the plane AB." Instead of saying "the vector sum of the lines $0A$ and $0B$ is the plane $0AB$," we can say "the vector sum of A and B is AB." The more we think of lines and subspaces as concepts in their own right, the easier it is to work with them.[16]

One analogy for this process is the notation used for long-distance telephone numbers. In the UK, every long-distance area code begins with a 0, for example the area code for Newcastle is 0191, and the prefix 0 is published as part of the code. However, if you dial to the UK from an international location, once you've dialled the country code (+44), you leave the 0 off from the beginning of the area code. This leads to confusing notation — for example, you might see a number on a business card written as (+44)(0)191 etc. which can be difficult. On the other hand, in the USA, long-distance calls are prefixed with a 1, but this digit is *not* usually written down explicitly — it's just common knowledge that you should dial 1 for long distance. Thus the code for San Francisco is not described as 1415, it's just 415. This means that internationally, a number can just be written as +1 415 etc., where the +1 is now a country code.

If we remove the unnecessary zeros from the beginning of our notation for lines and planes, we move from Euclidean geometry to *projective geometry*, so called because every point is naturally viewed from one point of view, the origin 0, just as if there was a camera placed at 0, free to rotate but unable to move to a different location.

If you only ever look at the space from the point of view of the origin 0, two points which lie on the same line through 0 will always appear to be in the same place — distance from 0 becomes unmeasurable. (This is a fairly practical model for the earth's view of the stars, consistent with the fact that astronomers throughout history have made excellent observations of the positions of the stars as seen from the earth, but opinions regarding the *distance* of the stars from the earth have been wildly inaccurate.) Thus, points which lie upon the same line become mingled together, so all points lying on the same line through the origin become representatives of the same 'meta-point' in projective space. The importance of projective geometry to applications such as perspective drawing was recognized early on by Desargues (1591–1661), though it was not until the work of Poncelet (1788–1867) that these methods became generally accessible, not long before Grassmann's period.

---

[16]The innovation of using a single letter to represent a line was in fact introduced by Descartes (1637, p. 5) at the very birth of analytic geometry.

In practice, if we make the decision to normalize all word vectors, our WORDSPACE behaves more like a projective space than a vector space. The availability of both vector spaces and projective spaces accounts for some terminological overlap — a *line* in a vector space corresponds to a *point* in projective space, a *plane* in a vector space corresponds to a *line* in projective space, and so on (see Coxeter 2003). This flexibility of terminology turns out to be very useful — many statements that had been proved separately for points, lines and planes turned out to be examples of the same constructions.

Grassmann (1862, §2) used the idea of identifying points on the same line, saying that two vectors a and b are *congruent* to one another if they are numerically related (i.e there exists a non-zero real number λ such that $a = \lambda b$ and $b = \frac{1}{\lambda}a$). The plot thickens — not only lines, but also planes and subspaces of any dimension can be thought of as 'points' in this space of possible subspaces. This 'metaspace' — the collection of subspaces of a vector space — is called a *Grassmannian manifold* to this day, and these spaces have become vital ingredients of differential and projective geometry, topology and theoretical physics (see Ward and Wells 1990, Ch 1). Thus projective spaces are the simplest kind of Grassmannian manifolds.

An important difference between Grassmannian manifolds and the vector spaces from which they are constructed is that Grassmannian manifolds are *compact*. For example, consider the set of lines passing through the origin in the plane, which is the simplest kind of projective space. Even though the plane itself is unbounded, every line can be represented by the point where the line crosses the unit circle, and so the set of possible lines is clearly bounded.[17] This goes part way to explaining how unbounded vector spaces stretching off to infinity in all directions be adapted to model the *limited* quanta described by Aristotle. The compactness of a space is very important for guaranteeing its tractability, because compact spaces can be 'covered' using a finite number of covering sets. For example, the earth's surface is compact, so

---

[17]The correspondence between lines in the plane and points on the circle is not exact, because each line crosses the unit circle not once but twice. Exact correspondences between spheres and projective spaces do occur in dimensions 2 and 4, since the 2-sphere S^2 is homeomorphic to the complex projective space $\mathbb{C}P^1$ and the 4-sphere S^4 is homeomorphic to the quaternionic projective space $\mathbb{H}P^1$. In general, the intuition that vector spaces are unbounded but spheres and projective spaces are compact is quite correct.

a spatial database can divide it into a finite number of cells. A spatial database can in principle divide the earth into 10 meter squares, and keep a record for each of these square cells without running out of memory. If the earth instead was an infinite plane, the database would be unable to keep a record for each square cell unless it had infinite memory.

7.5.2 Building higher-level concepts

For *any* point a on the line A and b on the line B, their sum $a + b$ is always somewhere in the plane containing A and B, so this plane is sometimes called the *vector sum* of the subspaces A and B. This sum operation can be generalized to subspaces of higher dimension — for any pair of subspaces A and B, their *linear sum* contains all points of the form $a + b$ where a can be any member of A and b can be any member of B. In this way, the concept of vector addition of *points* can be extended to the concept of vector addition of *subspaces*. An important intuition to grasp is that *the vector sum of two subspaces is the linear span of their set union*. For example, the set union of the lines A and B is a kind of cross-shape $A \cup B$. Their vector sum, the plane AB, is the linear span of this cross-shape (just as the plane of a map is the linear span of its eastward and northward axes). Since every point in this subspace is formed

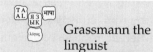

Grassmann the linguist

As well as being a great mathematician, Grassmann was a scholar of Indo-European language and literature, and his dictionary of Sanskrit, first published in 1873, is still available today.

Grassmann's Law, published in 1862, explained sound changes in Indo-European languages including Sanskrit and Greek. The law states that if consecutive syllables began with aspirated consonants (i.e. sounds which involve the release of a puff of air), then the first syllable lost its aspiration (for example, Greek *thriks*=hair becomes *trikhos* instead of *thrikos* in the genitive case). Grassmann's law challenged the once widespread belief that Sanskrit was *the* original Indo-European language, since Sanskrit had undergone changes like any other language. Grassmann's work, like Darwin's, was part of a new era of science which accepted the overwhelming empirical evidence that the world as we see it today is the result of gradual change (see Lehmann, 1993, §2.4).

from the sum of a point in A and a point in B, it is often written as $A + B$.

In Grassmann's algebra of conceptual structures in higher and higher dimensions, spaces of different dimensions are nested inside one another in a kind of lattice. This enables us to use lower-dimensional subspaces representing more specific concepts to be contained in higher-dimensional subspaces representing more general concepts. This brings some of the benefits of taxonomic classification, and it is a lot more flexible — whereas a node in a taxonomic tree has a unique path to the root, subspaces in vector spaces can trace their lineage via many different routes. For example, a line is contained in infinitely many different planes — thus, each of these planes could potentially be regarded as a 'parent concept' for the line. Grassmannian geometry naturally enables concepts to be cross-referenced.

As we move to higher dimensions, there are many more independent directions available to consider, which leads to much more diverse structures. For example, in a Euclidean (i.e. flat) space of three dimensions, the lines orthogonal to a given plane are all parallel to one another. This is a severe restriction — it guarantees, for example, that the line C in Figure 7.4 is the *unique* line through the origin perpendicular to the plane AB. Any query vector orthogonal to both A and B has only one choice left, C. If we have three dimensions in total, and remove two dimensions from the space of possible options, this leaves only one dimension free to vary. This leads to way too much dependency between variables to be able to sensitively model most interesting datasets. In higher dimensions the options are not nearly so restricted. For example, in a 100-dimensional WORDSPACE, if the components in any two-dimensional plane are removed from a vector, this still leaves 98 dimensions which between them may represent a great diversity of information.

Our WORDSPACE is precisely such an example — using singular value decomposition (page 175) to reduce our 1000 column-labels to 100 latent variables we project each word-vector from a 1000-dimensional space onto a 100-dimensional *subspace*. This is because in 1000 dimensions there is often *not enough* dependency between different directions — as opposed to three dimensions, where things are often too constrained, in 1000 dimensions there is so much space that information dissipates out in a very sparse cosmos. To counter this, we work in a more concentrated 100-dimensional subspace.[18]

As we have seen, adapting the concept of angles between lines to the higher-dimensional concept of the inclination between a line and a plane, or the inclination between two planes, is a process which is described in Book XI of Euclid's *Elements*. To further extend these concepts to give a general measure of similarity between subspaces of any dimension, we need turn no further than the works of Grassmann, who defines a system of product operations on extended quantities, from which the scalar product or *inner product* can easily be derived.[19] This abstract approach leads to the following general definition:

> Two quantities different from zero are said to be orthogonal if their inner product is zero. (Grassmann, 1862, §152)

The mathematical formation of these operations is not especially difficult (though for historical reasons, it is unfortunately rarely taught at an undergraduate level). The interested reader can turn to Grassmann's own exposition, and the excerpts of his work given by Smith (1929, pp. 684–696) are particularly well-chosen for those who wish to follow this conceptual story. The practical conclusion, for our purposes, is that the mathematics needed in order to remove two unwanted concepts from a query was written down by Euclid, and the mathematics needed to remove more than two unwanted concepts was written down by Grassmann.

7.6 Quantum Logic and beyond

How did Grassmann's nested collection of subspaces of a vector space come to be regarded as a *logic*, and in particular, why in 1935 did Birkhoff and von Neumann propose this structure as *The Logic of Quantum Mechanics*?

The underlying reason is to do with linearity and the way solutions can be superimposed upon one another. If the quantum state $|A\rangle$ is a possible solution for an equation involving these operators, it follows

---

[18]The choice of 100 dimensions has evolved over time on the Infomap project. For experiments comparing different numbers of latent dimensions see Landauer and Dumais (1997).

[19]A key step in this process is Grassmann's combinatory product, which is called the *exterior product* in modern algebra and differential geometry (see Lang, 2002, Ch. XIX).

that any state $\lambda|A\rangle$ will also be a correct solution. Moreover, if $|A\rangle$ and $|B\rangle$ are two independent solutions, then any state of the form $\lambda|A\rangle + \mu|B\rangle$ will also be a correct solution. This is exactly what Grassmann proposed — the solution set contains every state-vector that can be 'numerically derived' from the pure states, and so this solution set is precisely the subspace generated by these pure states. It follows that every 'experimental proposition' in a quantum-mechanical system corresponds to a subspace of the vector space in which the states of the system are represented mathematically (Birkhoff and von Neumann, 1936, §6).

7.6.1 Quantum disjunction

If a particle can be in the state $|A\rangle$ or $|B\rangle$, it can also be in a combined state $\lambda|A\rangle + \mu|B\rangle$.[20] Thus the disjunction in quantum mechanics is not given by the Boolean set union of $|A\rangle$ and $|B\rangle$ — it is given by the linear span of this union.

Definition 18 The collection of combined states of the form $\alpha_1|A\rangle + \alpha_2|B\rangle$ is called the *quantum disjunction* of the states $|A\rangle$ or $|B\rangle$.

The quantum disjunction is thus more general than the Boolean disjunction — it contains the elements of the Boolean disjunction (the pure states $|A\rangle$ and $|B\rangle$), and the combined states arising from these two — the elements in between which are necessary to smooth the discrete Boolean *set* into a complete vector *subspace*.

7.6.2 Quantum negation and the De Morgan Law

Many theorems in classical Boolean logic have exact analogues in quantum logic, and symmetries of this kind finally led to the Axioms of Logic being systematized. This material will be covered in more detail in Section 8.4, once we have developed the necessary lattice theory. The method of vector negation using orthogonal projection turns out to be exactly the right notion to be compatible with the quantum disjunction defined above.

---

[20]The numbers λ and μ might need to satisfy some conditions to ensure conservation of energy, mass, or probability — for example, normalizing the state vector by demanding that $\lambda^2 + \mu^2 = 1$ is a way of guaranteeing that the probabilities of observing the particle in different states add up to 1.

Logic with Vectors: Ambiguous Words and Quantum States / 239

One important result of this is that the De Morgan Law (*NOT* a) *AND* (*NOT* b) = *NOT* (a *OR* b) (equation 7.17, page 206) translates perfectly to quantum logic, and this gives a good account for why our vector negation operator performs so well when trying to remove *several* unwanted concepts.

If a user specifies that they want documents related to a but not $b_1, b_2, \ldots, b_n$, then (unless otherwise stated) it is normally clear that they only want documents related to *none* of the unwanted terms b_i (rather than, say, the average of these terms). Thus we want a to be orthogonal to *each* of the negated arguments, rather than (for example) the average of these arguments.

But if a is orthogonal to each of the b_i, it is orthogonal to the whole subspace generated by these vectors, and vice versa (if a is orthogonal to a generating set for the subspace, it is orthogonal to the whole subspace). In order to successfully create a query that is orthogonal to *all* of the unwanted keywords, the unwanted keywords *must* be thought of as generators of a subspace, not just as isolated points.

The query statement we began with is of the form

$$a \text{ } \textit{AND NOT } b_1 \textit{ AND NOT } b_2 \ldots \textit{ AND NOT } b_n$$

which according to the De Morgan Law can be translated into the statement

$$a \textit{ NOT } (b_1 \textit{ OR} \ldots \textit{OR } b_n). \tag{7.21}$$

In order to make this work in practice, we had to make the vector a orthogonal to the subspace spanned by all of the unwanted keywords b_i. Thus the expression $(b_1 \textit{OR} \ldots \textit{OR} b_n)$ being negated in equation (7.21) corresponds to this spanning subspace, precisely as predicted by quantum logic.

To check that this interpretation is of practical value, I carried out document-retrieval experiments (again, using queries of the form "keyword a *NOT* keywords b_1, b_2") which compared the method of making the query vector orthogonal to the entire plane generated by the unwanted keywords with the simpler technique of just negating the *average* of the unwanted vectors. There was a big difference between these — the method that just removed the average of the unwanted vectors did not do very well at removing either of the unwanted keywords from retrieved documents, whereas the method of making the query vector orthogonal to the whole unwanted plane did much better at removing *both* keywords. These results are also presented more thoroughly in Widdows (2003b), and they demonstrate that the theoretical correctness of the quantum-logical operations in WORDSPACE gives rise to good performance in practice, compared with more elementary vector methods which deal only with points and not with subspaces.

Quantum logic is a natural part of the Grassmannian algebra of vector spaces and their subspaces, and we will see in Chapter 8 that this is a fertile source of examples in lattice theory. Some philosophers (in particular Putnam 1976) have championed quantum logic, partly because it can provide some elegant reasons for the difference between classical physics and quantum mechanics. However, Grassmann's model does not depend on (and predates) its application to quantum mechanics, and it would be a mistake to tie the notion of 'vector logic' too closely to its interpretation as a logic for quantum mechanics. The operators we have described are available in any scientific model that uses vector spaces to represent information.

7.6.3 Disjunctions and the distributive law

The behaviour of the disjunction operator *OR* is one of the main differences between Boolean logic and other frameworks such as quantum logic. In all of these logical systems, a disjunction is an operator which takes two (or more) arguments and builds a more general concept or proposition out of these arguments. Several such operators have been described in this chapter, starting with 'CTRL + click' and 'SHIFT + click' ways of selecting multiple files or folders in Figure 7.1.

This difference goes back to whether the disjunction satisfies *distributive law* introduced by Boole (1847, Ch. 1) in the form

$$x(u + v) = xu + xv.$$

Here, Boole is asserting that his algebra of concepts behaves like the normal algebra of real numbers (for which this equation is always true). This enables a class to be assembled piecemeal by finding the u and v ingredients and combining the results. This process is assumed to be realistic in classical physics, but it cannot account for phenomena such as the interference observed in the double-slit experiment (page 219). In

terms of sets, the distributive law asserts that
$$A \cap (B \cup C) = (A \cap B) \cup (A \cap C),$$
so a point a cannot be in the set $B \cup C$ unless it is contained in at least one of the arguments B or C. While this is true in Boolean logic and set theory (which uses the 'CTRL + click' form of disjunction), the assumption fails for the 'SHIFT + click' form of disjunction, which may be used to select lots of folders or files without clicking on them explicitly.

The disjunction obtained using 'SHIFT + click' is usually a form of *bounding-box disjunction*, as described in the box on page 225. This is particularly easy to implement in Cartesian coordinates: if the opposite corners of a box have coordinates (x_1, y_1) and (x_2, y_2) then a third point (x, y) is inside the box if x is in between x_1 and x_2 and y is in between y_1 and y_2. This construction clearly does not obey the distributive law, since many points (x, y) can be included which are different from both (x_1, y_1) and (x_2, y_2). Since the non-distributive form contains more points than the distributive form, we may say that the non-distributive bounding-box disjunction is *more general* than the Boolean, distributive version.

Notice that the quantum disjunction also is more general than the Boolean disjunction. The Boolean disjunction of the states $|A\rangle$ and $|B\rangle$ corresponds to their set union $\{|A\rangle, |B\rangle\}$, which contains *only the pure states*. Their quantum disjunction, on the other hand, *includes the combined states as well*. The situation in WORDSPACE shown in Figure 7.3 is exactly the same kind of representation, where w_1 and w_2 represent the pure senses of the ambiguous word w, which lies somewhere on the line joining these two (once we impose the normalizing condition $||w|| = 1$, which is also part of the process in quantum mechanics).

Mathematically speaking, the difference between Boolean disjunction and quantum disjunction is the same as the difference between *subsets* of a set X and *subspaces* of a vector space V. The set union of two subspaces is not in general a subspace — the two lines A and B in Figure 7.4 form a cross-shape, which only becomes an actual subspace when the points lying evenly between A and B are added to turn the cross-shape into a surface. The act of adding all the points in between these lines is sometimes called *closure*, and the concept of closure can be applied to other geometric properties such as convexity. The variety of non-distributive disjunctions which arise from these different geometric closure conditions is explored by Widdows and Higgins (2004), who also describe the relationship between non-distributivity and the age-old philosophical problem of induction.

7.6.4 Different disjunctions in logic and language

Since there are a variety of mathematical operations that can play the role of an *OR* operator, it comes as no surprise that in everyday language there are different ways of interpreting statements involving disjunctions which are appropriate in different contexts. Boole himself assumes that disjunctions should be interpreted exclusively: for example, when considering the phrase

> things productive of pleasure or preventive of pain

in Mr Senior's definition of *wealth*, Boole says that

> it is plain from the nature of the subject that the definition precludes the notion of anything being both productive of pleasure and preventive of pain at the same time.   Boole (1854, Ch. 4)

(From this we may be tempted to infer that Boole was not a smoker.) This form of disjunction — either A or B but not both — is often called the *exclusive* OR or *XOR* connective.

Part of Boole's uneasiness here may arise from the problem of interpreting the symbol $p + r$ in his binary system, which only permits the values 0 and 1. The expressions pr for conjunction and $1 - p$ for negation work perfectly for both binary arithmetic and logic (as is confirmed by considering the truth tables in Figure 7.1). On the other hand, using the symbol $r + p$ for disjunction seems to break down in the case when r and p are both equal to 1, because this gives the result that $r + p = 2$ which is not a permissible value in Boole's system. His suspicions were correctly aroused: in order to do binary addition and multiplication consistently, it makes much more sense to say that $1 + 1 = 0$ than to that $1 + 1 = 1$.

Both the inclusive and exclusive disjunctions in classical logic are operations tied to the Boolean values of 0 and 1, which are discrete. For situations describing continuous quantities such as time and space, the quantum disjunction appears to be a more satisfactory model for what speakers intend. For example, if you were asking how far you had to travel to a given town and were told by a passer by

> The town is 10 or 15 miles away.

you would probably interpret this as meaning that the distance to the town is somewhere *in the region* of 10 to 15 miles: if you arrived at the town after travelling 11, 12, 13 or 14 miles instead, you would not think that the directions had been strange or misleading. Here the disjunction "10 or 15" is being interpreted as a quantum disjunction.

However, suppose that a discrete set of important points in space or time had already been picked out, as happens with a train timetable. If you were planning to catch a train and were told

> Suitable trains leave at 12 o'clock or 2 o'clock.

you would be more likely to think that these are the *only* trains between these times. This difference is affected by whether the measurement you are given is supposed to be continuous or discrete. In the first example, the possible distances from where you are standing to the town are continuous, and a quantum interpretation is appropriate. In the second, you presume that the set of trains, and hence the set of times at which they leave, is discrete, in which case a Boolean interpretation that does not permit in-between values or combined states is appropriate.

Conclusion

This chapter has demonstrated that, just as we can talk about different kinds of spaces and different geometries, we can also describe different logics. Like geometry, logic should not be regarded as a set of rules cast in stone and identical in all situations. We interpret sentences involving logical connectives differently in different contexts, sometimes depending on whether the space of possibilities is thought of as continuous or discrete.

The mathematical theories which enable these logical systems were developed in the mid-1900s. Boole's system was applauded in his day and has become the backbone of digital computing. Grassmann's system was largely ignored, and though it has been used successfully in many branches of physics and mathematics and laid the foundation for quantum logic, its full value is only now being recognized.

These different logics have tangibly different effects when used to process search queries in WORDSPACE. To remove a specific word or phrase from search results while remaining in more-or-less the same topical area, Boolean negation is better. To remove a word *and* move to a different topic, quantum negation is better (though to my knowledge it is currently only available in the Infomap WORDSPACE demonstration). Far from being a complicated and esoteric alternative to Boolean search, quantum-logical operations are simple to implement and completely natural in the right sort of models — which are already used to represent all sorts of data in many areas of science.

Wider Reading

This chapter began with a discussion of the contributions made by George Boole, which in particular are gathered in his 1847 work, *The Mathematical Analysis of Logic*, and his 1854 work *The Laws of Thought*. Both are wonderful to read: though the earlier work is sometimes described as preliminary and superseded by the later, they are very different in character. While *The Laws of Thought* is a mature Victorian treatise on a par with Darwin's *Origin of Species*, *The Mathematical Analysis of Logic* has more of the flavour of a concise and exact set of lecture notes for the advanced student, and many of Boole's mathematical contributions (especially to set theory) are particularly clear in this form.

Boolean logic and Boolean algebras are described by Hamilton (1982) and Davey and Priestley (1990); the use of truth tables in logic is covered in detail by Partee et al. (1993) and Gamut (1991); and the Boolean model for information retrieval is described by Baeza-Yates and Ribiero-Neto (1999).

Some of the different interpretations available for disjunctions in language are well-known and have been discussed by Boole himself and more recently in a standard and readable text on logic and language (Gamut, 1991, Ch 6). One of the most interesting discussions of various forms of negation is contained in Aristotle's treatise *On Interpretation*. For analysis of the difference between the negations in "I don't know" and "I don't want" (one involves lack of knowledge, whereas the other usually involves an actively opposite desire), see Wierzbicka (1996, Ch 1). For the story of negation through the ages, see Horn (2001).

Quantum theory is of course an enormous and fascinating topic. Classic formal treatments include those of Dirac (1930), von Neumann (1932) and Bohm (1951); for the reader seeking a more gentle introduction, the popular versions presented by Guillemin (1968), Polkinghorne (2002) and Squires (1986) are all excellent.

The original paper on quantum logic by Birkhoff and von Neumann (1936) gives a very clear description, not only of quantum logic itself but of the some of the foundations of lattice theory we will use in the next chapter. A more recent and thorough theoretical treatment is given by Cohen (1989), which gives a particularly lucid account of the algebra of projection operators. The role of quantum logic as a probability calculus is described by Wilce (2003), and some of the philosophical issues are discussed by Putnam (1976). Finally, *Geometry of Quantum Theory* (Varadarajan, 1985) is difficult but beautiful, giving a thorough account of the link between classical physics and Boolean logic, and motivating the quantum logic of vector spaces from some of the most standard and general algebraic considerations. If you are an expert reader of mathematics, Varadarajan's book is a challenge and a delight.

The adaptation of the work of Grassmann and Clifford to logic has been simplified in this chapter, and it is important to remember that a line or a plane (unlike a set) always comes with a direction or more generally an *orientation*. The framework of geometric algebra can thus be used to represent directed concepts such as journeys, and for this reason geometric algebra may be a better model than Boolean logic for supporting the representation of temporal concepts. The potential of geometric algebra to provide a unified model for physics and engineering is described by Lasenby et al. (2000) in an accessible paper that considers applications from special relativity and quantum mechanics to robotics and motion analysis.

8

Concept Lattices: Binding Everything Loosely Together

This is the final chapter in our story, and it explains how many of the mathematical systems we have encountered (hierarchies, vector spaces, and logic) can be viewed as special examples of a space called a lattice.

Much of the mathematics used in this book so far been rooted in developments during a critical period in the middle of the nineteenth century. Chapter 3 described the taxonomic approach to classifying concepts, through the work of Charles Darwin and beyond to the towering figure of Aristotle. The chapters on vectors introduced the geometric breakthroughs of Hamilton and Grassmann, themselves influenced by the work of Descartes. Chapter 7 also described the logic of Boole, which owes much both to Aristotelian logic and Cartesian algebra and geometry. These contributions, sharing many of the same roots, have enabled us to explore concepts using several exciting and diverse methods.

This process of branching out is very important and healthy — ideas that germinate in one field often bring insight, provoke questions and take on new form in completely different fields. But diversification can also lead to fragmentation, and some of the most important scientific advances have happened when scholars realized that ideas couched in very different terminology could actually be described by a single unified model. The analytic geometry of Descartes (Section 5.3), for example, enabled problems in geometry (such as finding the points where lines and curves intersect) to be solved using methods from algebra (by describing said lines and curves by equations representing the relationships between their x and y coordinates, and solving these equations to find the numerical coordinates). Broadening this unification into the physical sciences, Isaac Newton went on to represent *time* as a coordinate, using *equations of motion* to predict where a particle would be as a function of time. The accurate laws that had been formulated by Galileo Galilei (1564–1642) in the earthly realm and by Johannes Kepler (1571–1630) in the heavenly spheres could then be understood as manifestations of the same *universal law of gravitation*, which states that every object attracts every other object.[1] More recently, the apparent incompatibility between quantum theory and general relativity (relativity is local, quantum theory non-local; relativity is deterministic, quantum theory non-deterministic) has led physicists beginning with Einstein to search for an underlying Grand Unified Theory.

Our final chapter is about a distinctly twentieth century branch of mathematics[2] that weaves our previous threads together, treating the hierarchies of Aristotle and Darwin, the geometry of Hamilton and Grassmann, and the logic of Boole as different variants of one underlying structure called a *lattice*. In a way, this is fitting example for the historical process of divergence and convergence we have described: the distinctive property of a lattice is that any two elements can be *disjoined* (allowed to spread out and diversify) and *conjoined* (woven together).

All this talk of twentieth-century mathematics and unifying theories should not put you off: lattices are not particularly complicated, and like many of the newest branches of scholarship, the ideas behind lattice theory go back at least to classical Greece, and beyond this to roots in astrology and mysticism lost beyond antiquity.

We will navigate through this chapter by considering the different

---

[1] This way of describing gravitation is distinctly pre-relativistic — Einstein's theory does not describe massive objects pulling at one another, but rather massive objects curving the space around them so that the natural path or *geodesic* of other objects bends with least effort towards them. However, the main tenet of Newtonian gravity, that every object exerts gravitational influence over every other object, purely on account of its mass, irrespective of what the objects are made of, remains a cornerstone of physics, and would have astonished many of the ancient philosophers.

[2] As we shall see, however, Grassmann wrote down the fundamental principles of lattice theory in 1862 — had his work been properly considered by professional mathematicians at the time, the development of algebra may have been accelerated by several decades.

ingredients of a lattice in turn, starting with ordered sets (Section 8.1), examples of which include numbers, trees or hierarchies, and some parts of Aristotelean logic. Section 8.2 explains the *join* and *meet* operations that make an ordered set into a *lattice*, examples of which include numbers, lattices of sets, and vector spaces. Section 8.3 introduces the theory of *Concept Lattices* (Ganter and Wille, 1999), which can be used to describe both the way Aristotle portrays the four elements and the way a modern interlingua (that of Janssen 2002) represents words for kinds of *horse*, both applications giving strikingly similar models. Other examples show how the lattice operations give the lowest common multiple and highest common factors in whole number lattices, and eventually the disjunction (**OR**) and conjunction (**AND**) operations in logic. Finally, Section 8.4 surveys some of the opportunities for defining a suitable negation or *complementation* operation on a lattice, again using vector spaces and set theory as examples. This property distinguishes those lattices which can be thought of as *logics*, the last kind of space we will define axiomatically in this book.

8.1 Ordered Sets and Implication

Ordered sets form the most basic building block at the beginning of the journey to lattices and logics — and many other important mathematical structures. In fact, we have already described the key properties of an ordered set in previous chapters, but we haven't collected them together formally. Hopefully many of the examples presented below will therefore be familiar.

8.1.1 The real numbers, ordered by size

The real numbers $\mathbb{R}$ (and hence all of their important subsets, the rationals $\mathbb{Q}$, the integers $\mathbb{Z}$, and the natural numbers $\mathbb{N}$) have a natural ordering relationship given by their size. This relationship is written using the symbol $\leq$. For any two real numbers (or fractions, or integers, or natural numbers) x and y, we write $x \leq y$ if and only if x is less than or equal to y, so for example $2 \leq 3$, $2 \leq 5$, and also $2 \leq 2$.

For three numbers x, y and z, if $x \leq y$ and $y \leq z$, we also know that $x \leq z$ (for example, if $x = 2$ and $y = 3$, if the number z is bigger than or equal to 3 we know that it must also be bigger than or equal to 2). As we saw in Section 4.4, this property can be summed up by saying that the relation $\leq$ is *transitive*.

Another important property of the relation $\leq$ is that for two numbers x and y, the statements $x \leq y$ and $y \leq z$ can both be true at the same time *only* if x and y are the same number. This is summed up by saying that the relation $\leq$ is *antisymmetric* (Section 3.3).

Finally, any number is less than or equal to itself (because it's equal to itself), so we have the rather trivial property that $x \leq x$ for all numbers x. This property is summed up by saying that the relation $\leq$ is *reflexive*.

These vital properties enable us to use the real numbers so successfully for making measurements. As described in Section 4.4, many words in language that compare the magnitudes of two properties (such as **higher, older, hotter**) have similar properties to those exhibited by the real numbers ordered by $\leq$ (for example, if A is older than B and B is older than C, it follows that A is also older than C).[3]

In describing physical properties such as **altitude, age** and **temperature** using measurements such as **height, years** and **degrees celcius** (as in Table 1.1), the different possible states of the system inherit the *ordering* $\leq$ from the (subset of) real numbers to which they are mapped. For a measuring process and apparatus to be acceptable, it must always map higher states to bigger numbers and lower states to smaller numbers — in other words, to be a good measurement, a mapping from the states of a system to the real numbers must be an *order-preserving map*. For example, if alcohol and mercury expanded sporadically, unreliably and reversibly as their temperature increased (instead of gradually, reliably and above all, *monotonically*), the thermometers invented by Daniel Fahrenheit (1686–1736) would not have preserved the natural ordering on temperatures and would have been useless.

The properties of reflexivity, transitivity and antisymmetry are the defining properties of an *ordered set* (Figure 8.1). We will use the symbol $\preceq$ rather than $\leq$ for a general ordering relationship, because $\leq$ has its specific meaning in the context of comparing the sizes of two numbers, and as we'll see later this could be confusing when we're talking about

---

[3]Most such comparatives in English follow a pattern more similar to the mathematical relation $<$ (is less than) rather than the relation $\leq$ (is less than or equal to). The former notion is more intuitive, the latter is often regarded as more fundamental in mathematics: the difference is much less than a hair's breadth, at least in continuous models.

> **Formal Definition of an Ordered Set**
>
> An *ordered set* is a set P with a relation on $\preceq$ on P such that
>
> - O1. For all $x \in P$, $x \preceq x$. This is called *reflexivity*.
> - O2. For all $x, y \in P$, $x \preceq y$ and $y \preceq z$ imply that $x = y$. This is called *antisymmetry*.
> - O3. For all $x, y, z \in P$, $x \preceq y$ and $y \preceq z$ imply that $x \preceq z$. This is called *transitivity*.

FIGURE 8.1 An ordered set must satisfy these three axioms.

other ordering relationships between numbers which are, if anything, even more interesting.

8.1.2 Implication, containment and hyponymy

Many other relationships between concepts and statements, in particular those of those of *implication*, *containment* and *hyponymy*, also satisfy the ordered-set axioms. As we have already seen, these three relationships are closely linked to one another: the statements

- *horse* is a hyponym of *mammal*.
- The set of *horses* is contained in the set of *mammals*.
- x is a *horse* implies that x is a *mammal*.

are all really equivalent. The relationships in these statements all have special symbols of their own, and using notation introduced in previous chapters, we can write these statements as follows:

horse $\sqsubseteq$ *mammal*	*horse* is a hyponym of *mammal*
$\{horses\} \subseteq \{mammals\}$	The set of *horses* is contained in the set of *mammals*
$x \in \{horses\} \Rightarrow x \in \{mammals\}$	x is a *horse* implies that x is a *mammal*

These equivalences are well-known — the opportunity to use equivalent systems to talk about classes or sets and about propositions in logic was grasped by Boole (1847, Ch 5) from the outset. All these relationships obey the axioms in Figure 8.1: it follows that the concepts in a taxonomy, the collection of subsets of a given set, and the collection of statements in logic, all form partially ordered sets.

8.1.3 Directed Acyclic Graphs and Ordered Sets

The ordered set axioms in Definition 8.1 can be used to guarantee that there is a convenient and intuitive way to draw a geometric picture of an ordered set S, and this picture is sometimes called a *Hasse diagram*.

The technique works by drawing a small circle or node for each element of S, and drawing a line between two nodes x and y if $x \preceq y$. The direction of this relationship can be correctly inferred if we make sure that y is above x in the diagram. We have noted that the relationship IS_ABOVE is transitive — if A is above B and B is above C, then A is certainly above C. It is also antisymmetric — if A is above B then B can't also be above A. It follows that the transitivity and antisymmetry properties, critical to the definition of an ordered set, will be reflected in the Hasse diagram.

As an example, consider the following ordered set S, a subset of the Tree of Life (Section 3.1). Suppose that the set S has the elements

{ dog, dolphin, frog, bee, jellyfish, mammal, vertebrate, invertebrate, animal },

and that an ordering of S is given by the relationships

dog $\sqsubseteq$ mammal, dolphin $\sqsubseteq$ mammal, mammal $\sqsubseteq$ vertebrate, frog $\sqsubseteq$ vertebrate, bee $\sqsubseteq$ invertebrate, jellyfish $\sqsubseteq$ invertebrate, vertebrate $\sqsubseteq$ animal, invertebrate $\sqsubseteq$ animal.

We also infer the longer-range relationships *dog $\sqsubseteq$ vertebrate, dog $\sqsubseteq$ animal*, etc. because of the transitive assumption. The Hasse diagram of this ordered set is given in Figure 8.2. Just as with our first example of a simple graph (Figure 2.1), a simple exercise is to check that the set and the relations above and the diagram in Figure 8.2 are effectively equivalent. The relationship $x \sqsubseteq y$ can be read from the graph if there is an upward link from x to y, or a chain of links. Just as we don't bother to write *dog $\sqsubseteq$ vertebrate*, but assume it from the relationships *dog $\sqsubseteq$ mammal* and *mammal $\sqsubseteq$ vertebrate*, we don't bother to draw the link from *dog* to *vertebrate*, since it can easily be traced in the diagram and drawing *all* of the links would make the diagram tiresome to draw and pretty cluttered to read.

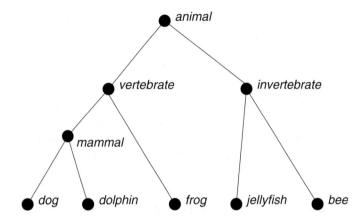

FIGURE 8.2 Hasse diagram showing several animals and their phylogenetic ordering.

So far, this example may have left you with the question "Isn't this just a diagram of a tree like all the ones in Chapter 3?" In this case, of course, this is correct — and in fact, *every* tree or hierarchy is a kind of ordered set. Much of this chapter is about the way higher mathematics fuses together different ideas into a harmonious whole, and as we go deeper into lattice theory we will find more of our paths meeting.

However, a tree or hierarchy is a very special sort of ordered set — namely, one in which each node has a *unique* path from itself up to a single root node. This entails that each node has just a single parent node. There are many ordered sets where this is not the case — for example, suppose we add a bit of extra world knowledge to our ordered set of animals, saying that a *dog* lives on land, a *bee* lives on land and in the air, a *dolphin* and a *jellyfish* live in water, and that a *frog* lives on land and in water. This gives the diagram in Figure 8.3.

This structure isn't a tree any more — it branches out in both upward and downward directions, and there are four nodes that don't have any parent nodes, rather than a single root node. However, the graph certainly represents an ordered set, as the new relationships are still consistent with the Axioms in Figure 8.1.

In talking about nodes and links, we are clearly using the language of graph theory as introduced in Chapters 2 and 3, and it's easy to see that the Hasse diagram representing any ordered set will be a kind of

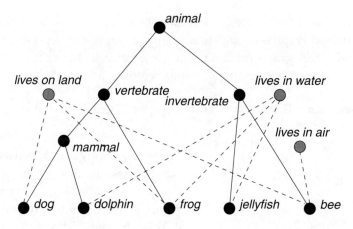

FIGURE 8.3 Hasse diagram showing animals, their phylogenetic ordering and habitats.

directed graph (as in Section 3.2), where the direction of the relationships is indicated by placing one node higher than another. But the Axioms in 8.1 place extra conditions on this directed graph — since $x \preceq y$ and $y \preceq x$ are incompatible when x and y are different, the small two-node directed graph in Figure 8.4 is ruled out. (Another way to see this is to note that we couldn't draw this graph as a Hasse diagram where the direction of the links is given by the relative heights of the two nodes.) Such a closed loop in a graph is often called a *cycle*, and a closed loop with just two nodes is called a *2-cycle*. This graph cannot represent an ordered set, and also any graph that contains a 2-cycle like this cannot represent an ordered set, since the axioms in Figure 8.1 have to hold for *all* of the elements of the set, and so even if the axioms are violated by just these two elements x and y, the structure still breaks down.

In fact, this problem applies more generally than just to 2-cycles. Recall that, because of transitivity, if there is a chain $x_1 \preceq x_2 \ldots \preceq x_n$ in an ordered set, then $x_1 \preceq x_n$ must also be true, and even if we don't draw this link explicitly in our diagram we still regard it as being there. (See the dotted link in Figure 8.5.) If there is also a link $x_n \preceq x_1$, then we have a 2-cycle just like in Figure 8.4, so

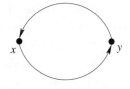

FIGURE 8.4 A 2-cycle in a directed graph.

this graph also can't represent an ordered set. We conclude that *any* cycle in a directed graph is enough to prevent the graph from representing an ordered set. A graph containing no cycles in it at all is called *acyclic*, which leads to the following important correspondence:

Observation 2 Every ordered set corresponds to a *directed acyclic graph*.

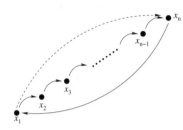

FIGURE 8.5 A general cycle in a directed graph.

This draws an important and interesting equivalence between a class of essentially *algebraic* objects (namely, sets satisfying the axioms in Figure 8.1) and essentially *geometric* objects (namely, directed graphs not containing any cycles). It's also interesting that so much of this correspondence can be understood by grasping the mathematical properties of a relationship like IS_ABOVE, which enables us to draw any ordered set as a Hasse diagram.

8.1.4 Hierarchies, disjunction and class labelling

For any set of nodes in a hierarchy, the lowest node that subsumes all of them can be regarded as their *disjunction* or *join*, as it will later be called. For example, in Figure 8.2, the join of *dog* and *dolphin* is *mammal*, and the join of *dog* and *frog* is *vertebrate*. These join nodes can be described in the following sentences:

> If x is a *dog* or a *dolphin*, then x is certainly a *mammal*.
> If x is a *dog* or a *frog*, then x is certainly a *vertebrate*.

Even though other more general concepts such as *animal* could fulfill the role of the concept x in these statements, we have deliberately chosen the most specific concept possible that fulfills the stated requirements. This node corresponds to the disjunction (OR operator) in logic, because it is the most specific node *implied by* both of its input nodes. It is called the join in lattice theory, because it is the node where paths up from the different inputs to the root node join together.

Now, the idea of finding a single node that gives a good description of a whole set of words is the motivation for the *class labelling algorithms* of Section 3.6. Here again we discussed the trade-offs between being too specific (and potentially wrong) and too general (and potentially uninformative). We explicitly introduced the idea of using the lattice join operation as a class label in Example 5 (page 93), and rejected this idea as being too sensitive to outliers in the data. Though lattice joins and disjunctions are well-motivated theoretically, when working with empirical data it is often worthwhile to bend the theoretical rules a little. However, it would certainly be reasonable to describe the class labelling algorithms we used as *statistical* join operations: like the theoretical join, the class labelling algorithms we used try to find the most specific class label that accounts for as much of the data as possible, but for the sake of getting good, useful answers, we do not require that the best class label can account for *all* the data.

As we introduce more of the theoretical tools of lattice theory and logic (including conjunction and negation), it is important to realize that in practical natural language processing systems, these theoretical tools will often be idealized solutions that need to be bent slightly to cope with the astonishing richness and complexity of the real-world situation.

8.1.5 Aristotle's Logic and Ordered Sets

As an early example of the importance of ordered sets in logic, we will briefly summarize some parts of Aristotle's logic. The rules of Aristotle's logical system embody at least some of the formal properties which we nowadays sum up in the axioms for an ordered set. For example, the following two quotations refer to properties which today are described as transitivity:

> When one things is predicated of another, all that which is predicable of the predicate will be predicable also of the subject.
>
> (*Categories*, Book I Ch 3)

and even more explicitly:

> If A is predicated of all B, and B of all C, then A must be predicated of all C.   (*Prior Analytics*, Bk I Ch. 4)

This train of reasoning can be summed up in the form

> Every C is a B, and every B is an A: therefore every C is an A.

Such a structure is called a *syllogism*, and Aristotle's logic revolves around analyzing which of the possible syllogisms are true and which are not. The variety of possible syllogisms comes from studying *quantification* and *negation*.

Those who are familiar with basic logic ('first order predicate calculus') may already know about the two quantifiers *some* or *there exists* (written ∃) and *all* or *every* (written ∀) (Gamut, 1991, §3.2). These allow us to generate quantified syllogisms such as

> Some people are philosophers, and all people are mortal:
> therefore some philosophers are mortal.

Once we add negation to this system, we can generate even more syllogisms, such as

> All horses are mammals, and no mammals are invertebrates:
> therefore no horses are invertebrates.

The combination of universal, particular and negated statements gives four possible kinds of assertion: all A are B (universal affirmative), no A are B (universal negative), some A are B (particular affirmative) and some A are not B (particular negative).

Geometrically, these propositions and the way they are interrelated can be represented as a *Square of Opposition* (Figure 8.6). The top pair of propositions are *contrary* to one another, the bottom pair are *subcontrary* to one another, and diagonally opposite propositions are *contradictory* to one another, where these terms are defined as follows:

- Two propositions that cannot both be true are *contrary*.
- Two propositions that cannot both be false are *subcontraries*.
- Two propositions that cannot both be true *and* cannot both be false are called *contradictories* (we might also say *opposites*).

I have placed the downward vertical arrows at the sides of the diagram to show that the propositions above imply the propositions below. Traditionally these are called *subalterns*, which may be confusing since the statements below are more general rather than more specific. One way to resolve this confusion would be to draw the square the other way up — this would go against centuries of tradition in the teaching of logic, but would have the effect of making the diagram consistent with the more general principles of drawing Hasse diagrams for ordered sets, as described in the previous section.

This Square of Opposition has fallen into disrepute in modern times because it fails to take into account empty sets. For example, does it really make sense to say that "Every *unicorn* is *white* implies that some *unicorns* are *white*" when, as far as we are aware, there aren't any *unicorns* in the first place? (One might be surprised at the clamour that has

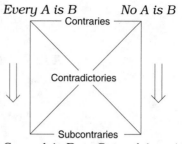

FIGURE 8.6 Square of Opposition.

been raised in modern times because Aristotle's logic doesn't cover such vacuous statements!) Another point to be aware of is that the four types of proposition involved in the Square of Opposition form only a small part of Aristotle's logical theory (described mainly in the works *On Interpretation* and *Prior Analytics*). Aristotle devotes at least as much attention to the effect of negation in different *parts* of the propositions, statements involving *individuals* (such as **Socrates**), and to statements involving conditionality and possibility (nowadays called different *modalities*). As often happens, later textbooks may give the impression that ancient scholars were narrow, but this misconception is often dispelled by reading the ancient scholars themselves.

To sum up, this section has hopefully given the reader a good flavour of what can be accomplished if we take a set of propositions, use the relation of implication to turn this into an *ordered* set of propositions, and add negation and quantification to the structure.

8.2 Lattices in Mathematics

Leaving aside quantification and (for now) negation, this section goes back to our basic ordered set structure and considers the effect of adding a complementary pair of operators that will correspond to disjunction and conjunction. These are the two key operators necessary for turning an ordered set into a *lattice*, and they are called the *join* and *meet* operations in lattice theory. Again, we have already come across these operations in a number of different guises, as will soon become clear. To begin with, let us consider a classic example of a lattice from number theory.

258 / GEOMETRY AND MEANING

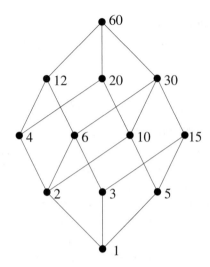

FIGURE 8.7 Lattice of the factors of the number 60.

The diagram in Figure 8.7 shows all the numbers that exactly divide the number 60 (that is to say, all the natural numbers x such that $\frac{60}{x}$ is a whole number). These numbers are called the *factors* of the number 60, and they comprise the set $\{1, 2, 3, 4, 5, 6, 10, 12, 15, 20, 30, 60\}$. This set can be given an interesting ordered set structure using the relationship 'is a factor of' or 'divides' so that we write $a \preceq b$ precisely when a is a factor of b (equivalently, b is a multiple of a). For example, we may write the relationships

$$2 \preceq 4 \quad \text{and} \quad 3 \preceq 15.$$

This is a more special relationship than just saying that (for example) $3 \leq 15$, because there are many other numbers that are *smaller* than 15, but that are not *factors* of 15.

Figure 8.7 shows the Hasse diagram of this ordered set. Have a good look at this diagram and convince yourself that (for example) the statement $2 \preceq 20$ (2 divides 20) can be read from the diagram correctly by tracing a path from the 2 node up to the 20 node.

So far, we clearly have an ordered set, and in its way Figure 8.7 is rather beautiful — much more interesting than just lining up the numbers from 1 to 60 in a ladder with bigger numbers at the top and smaller numbers at the bottom (the diagram which we would get from the ordering given by the usual $\leq$ relationship). It has a wonderful symmetry — if you rotate the diagram through 180 degrees, you will get an inverted copy of the same structure, but this time with the number 1 at the top and 60 at the bottom, and indeed, if you multiply any pair of numbers that are exchanged by this rotation (any pair of numbers that are opposite to one another in the diagram), their product is 60. You could draw a similar diagram taking any number instead of 60 as your top node and arranging its factors below it, but the structure is particularly rich for the number 60. This is one way to see why 60 was such a good choice as a base for the Babylonians to devise their number system and notation, as described in Section 1.3.

As with hierarchies and Aristotelean logic, the structure of this set allows us to infer new conclusions from chains of smaller observations. Suppose we know that a number a is a factor of 6, which we've written $a \preceq 6$. Then it follows that a is also a factor of 12, 30 and 60 (and of course many other numbers, but these are the ones in our diagram). We can get these results easily by transitivity in our ordered set, since (for example) $a \preceq 6$ and $6 \preceq 30$ must also imply that $a \preceq 30$. The same argument can be applied in the opposite direction, tracing paths downwards so that, for example, if we know that 12 is a factor of b for some number b, we can infer that 6, 4, 3, 2 and 1 are also factors of b by reading the diagram in Figure 8.7: since these numbers are all factors of 12, they must also be factors of the number b.

8.2.1 Meet and join in whole number lattices

Now, we take the crucial step from an ordered set to a lattice. Intuitively, the word *lattice* conjures up some sort of interlaced framework, such as a regular crystal structure, and to describe Figure 8.7 as a lattice looks quite intuitive. The critical property that makes it a lattice in mathematical terms is that the paths upwards from any two nodes always join together sooner or later, and the paths downwards from any two nodes always meet. We'll begin by defining these joining and meeting points, and then describe them more intuitively.

Definition 19 Join and meet

The *join* of two nodes a and b is their *least upper bound*, that is to say, the lowest node c such that $a \preceq c$ and $b \preceq c$.

Conversely, the *meet* of two nodes a and b is their *greatest lower bound*, the highest node d such that $d \preceq a$ and $d \preceq b$.

Join and meet nodes in number lattices are particularly interesting because they correspond to the lowest common multiple and highest common factor of two numbers (Figure 8.8). The *lowest common multiple* of two numbers is the smallest number which is divisible by both of them. For example, the lowest common multiple of 2 and 3 is 6, and the lowest common multiple of 10 and 12 is 60. We can use the Hasse diagram in Figure 8.7 to work out this information for us — follow paths up from 2 and 3, and you'll see that these paths join at the 6 node. Similarly, paths can be traced from 12 up to 60 and from 10 up to 60. In the case of 2 and 3, you can also trace paths up which join at 12 or which join at 30 — but these can be simplified to paths which join at the 6 node and carry on up together from there. Because of this, we say that 6 is the *join* of 2 and 3 in the diagram, and 60 is the join of 10 and 12. In the same way, we can follow paths downwards to find the *highest common factor* of two numbers. For example, paths down from 20 and 15 meet at the 5 node, which is the *highest common factor* of 15 and 20, and this node is called the *meet* of 15 and 20 in the diagram. The join of 2 and 3 and the meet of 15 and 20 are shown in Figure 8.8.[4]

Definition 20 Lattices

An ordered set in which every pair of elements has a unique, well-defined join and meet of this sort is called a *lattice*.

The join of two elements is written $a \vee b$ and the meet of two elements is written $a \wedge b$.

Not every ordered set is a lattice, because the join and meet of two elements might not exist, or might not be uniquely defined. For example, a standard family tree or *genealogy* that depicts ancestors and descendants in a family is *not* in general a lattice, because the least upper bound (in this case, the nearest common ancestor) of two people is not always uniquely defined (see the box on page 263). However, some of the most important spaces and sets we've met in this book *are* lattices, and the next few examples will explain why this is so.

---

[4] To remember the difference between join and meet, it can help to think that two points *join* to make a line, and two lines *meet* at a point. So the join is a broader more general concept, the meet is a narrower more specific one.

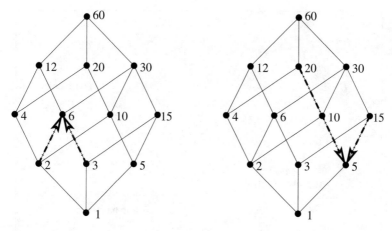

FIGURE 8.8 The join (lowest common multiple) of 2 and 3 is 6, and the meet (highest common factor) of 15 and 20 is 5.

8.2.2 The natural numbers and the harmonic series

We've seen that the natural numbers

$$\mathbb{N} = \{1, 2, 3, 4, \ldots\}$$

can be given the structure of an ordered set by saying that, for two numbers a and b, $a \preceq b$ if and only if a is a factor of b. We showed this mainly for the factors of the number 60, because this is a particularly beautiful example, but the principle can be applied to all of the natural numbers. With this ordered set, the least upper bound or join of a and b is their lowest common multiple, and the greatest lower bound or meet of a and b is their highest common factor. In symbols, we could write down the following examples:

$$6 \vee 40 = 120 \quad \text{and} \quad 6 \wedge 40 = 2.$$

As a practical example, you can actually *hear* this lattice structure in action if you have a guitar handy (it's a very quiet effect so an electric guitar with the volume turned up is the easiest way to demonstrate it). The demonstration works by playing the *harmonics* on a string, gently touching two places on the string at once. As described in Figure 1.4, if you touch the string very gently half way along (just above the twelfth fret), you produce a note one octave higher that the fundamental frequency, the 2nd harmonic. If you touch the string one-third of the way along (just above the seventh fret) you will hear the 3rd harmonic, one

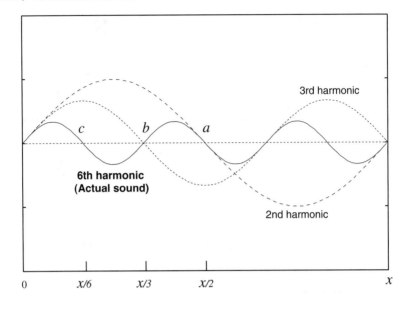

FIGURE 8.9 Higher harmonic waves on a string

and a half octaves above the fundamental frequency. These harmonics are shown again in Figure 8.9, a and b marking the places where you constrain the string to remain still, only allowing through the waves which leave the string stationary at this point.

So, what happens if you touch the string at *both* places, a and b? The only waves that can resonate along the string are those which leave both the points a and b undisturbed, and as you can see in Figure 8.9, the largest wave which is stationary at both of these points is the *6th* harmonic, which would be equivalent to touching the string gently at the point c (one-sixth of the way along, or just above the third fret), though in practice, the sound obtained by touching the string in the two places shown in the Figure 8.9 is normally slightly clearer and more satisfying. This is an elegant example of a situation where nature picks out a wave with both the twoness of the 2nd harmonic and the threeness of the 3rd, and the first example nature finds is the 6th harmonic, because 6 is the lowest common multiple of 2 and 3. In a very real sense, the guitar string is a naturally occurring lattice in which 2 and 3 make 6.

8.2.3 Set theory and lattices

Set theory gives us another very important class of lattices, which turn out to be intimately related (and in fact equivalent) to Boolean logic. Consider a collection of different sets. The set containment relationship $\subseteq$ gives an ordering on this collection (obeying the ordered set axioms in Figure 8.1). For example, if the set A is contained in the set B ($A \subseteq B$) and the set B is contained in the set C ($B \subseteq C$), then the set A is contained in the set C, and so on. To make this notion more concrete, we normally define a 'universal set' U and say that all our sets are subsets of U. Then the set of all subsets of U is itself an ordered set under the relationship $\subseteq$.

Now, recall the set union and intersection operations $\cup$ and $\cap$ (back in Figure 1.3). The set union $A \cup B$ contains every element of A and every element of B, and no elements that are in neither A nor B. It follows that $A \cup B$ is the *smallest* set that contains A and B — any other set containing both A and B must also contain $A \cup B$ (Figure 8.10). This is exactly equivalent to saying that $A \cup B$ is the least upper bound of A and B, with respect to the set containment relationship $\subseteq$.

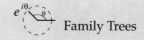 Family Trees

A very common example of an ordered set is a *family tree*, in which $a \preceq b$ when a is a descendant of b. However, a family tree is *not* really a 'tree' in the formal sense, because each person has more than one immediate parent, and there is no single root node (unless we all trace our family trees to the same primeval soup). A family tree is not in general a lattice, either, because the join of two nodes may not be uniquely defined — in the tangled but perfectly possible family tree below, *Iris* and *John* have two pairs of grandparents in common, so their nearest common ancestor is multiply defined.

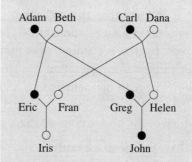

In bigger family trees, the chances of two people having many common ancestors who are unrelated to one another is quite likely.

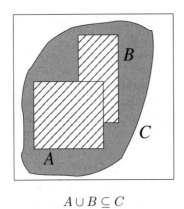
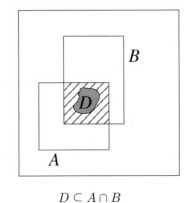

$A \cup B \subseteq C$ $D \subseteq A \cap B$

FIGURE 8.10 Every set containing both A and B contains $A \cup B$, and every set contained in both A and B is contained in $A \cap B$.

In the same way, the intersection $A \cap B$ of two sets A and B is the *largest* set that is contained in both A and B, since any other set that is contained in both A and B is contained in $A \cap B$. This is exactly equivalent to saying that $A \cap B$ is the greatest lower bound of A and B, with respect to the set containment relationship $\subseteq$ (also in Figure 8.10).

It follows that the collection of subsets of any set U, ordered by the containment relationship $\subset$, forms a lattice, with the join operation given by the set union operation $\cup$ and the meet operation by the set intersection operation $\cap$. These observations are often traced to the American mathematician, logician and philosopher Charles Pierce (1839–1914), who developed much of the notation used in modern logic. Much of the underlying set theory was developed by Boole (1847).

8.2.4 Vector space lattices

Our final example in this section is given by vector spaces and their subspaces. In Section 7.6.3, we described the way in which 1-dimensional lines or 2-dimensional planes can be thought of as vector spaces in their own right, which we call *subspaces* of 3-dimensional space. This gives us a very special ordered set — the total 3-dimensional space contains infinitely many distinct planes, and each plane contains infinitely many lines. We focus on lines and planes that go through the origin point 0, which guarantees that all the subspaces have at least one point in common (and enables us to use the techniques of projective geometry as introduced in Section 7.5.1). This forces the lattice to have a unique zero node at the bottom, which is subsumed by every other node in the lattice (i.e. $0 \preceq A$ for every subspace A).

It was also shown that the smallest subspace that contains two distinct lines through the origin is the plane which those lines generate (Figure 7.4). Conversely, the intersection of two distinct planes through the origin is the line common to both.

In other words, the set of lines and planes in 3-dimensional space forms a lattice, where the *join* of two lines is a plane, and the *meet* of two planes is a line. This example can be extended to any number of dimensions — the join of two subspaces A and B is their *vector sum* $A + B$ and the meet of two subspaces is their standard intersection $A \cap B$ (which is always itself a subspace).

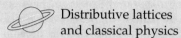 **Distributive lattices and classical physics**

The difference between classical and quantum physics is sometimes encapsulated by the associated lattice of experimentally verifiable observations, which in classical physics forms a distributive lattice and in quantum physics forms a non-distributive lattice. In a distributive lattice, the law

$$A \wedge (B \vee C) = (A \wedge B) \vee (A \wedge C)$$

always holds, which guarantees (for example) that the statements $a \in B$ and $a \in C$ can be measured independently and the results combined. In non-distributive lattices such as those in vector spaces, such a post-hoc combination is not justified, which explains why many physical quantities such as position and momentum are not simultaneously measurable.

Grassmann was well aware of these operations, and their description in terms of the ordering of concepts. In the very next paragraph after he gives the modern definition of the dimension of a vector space or subspace (translated as *domain*) he writes (Grassmann, 1862, §15):

> If every magnitude of a domain (A) is simultaneously a magnitude of another (B) (without the converse obtaining), then I call the two magnitudes *incident* upon one another, and say that the first domain (A) is *subordinate* to the second, the second *superordinate* to the first.
>
> **The collection of magnitudes that simultaneously belong to two or more domains are called their** *common* **domain, and the collection of magnitudes of two or more domains their** *covering* **domain.**

In one translation I have seen these words rendered as *meet* and *join*, emphasizing the connection with modern lattice theory. In case there was any doubt about this, Grassmann follows this definition with the following illustration:

> If for example the domain A is derivable from the units e_1, e_2, e_3 and domain B from the units e_2, e_3, e_4, then the common domain of the domains A and B is the domain derived from the units e_2, e_3, and the covering domain of A and B that derived from the units e_1, e_2, e_3, e_4.

No clearer example of the meet and join operations in a lattice of subspaces could be envisaged — there can be no doubt that Grassmann was familiar with the operations that eventually came to model the conjunction and disjunction in quantum logic. Astonishingly, to his great list of largely uncredited honours, we may add that Grassmann wrote down the fundamental principles of lattice theory way back in the mid-1800s.

8.2.5 Why talk about lattices?

In a book that is about geometry and *meaning*, some readers might have begun to wonder what this final chapter has to do with meaning in language at all, given so many mathematical examples. In the next section, this balance will finally be redressed, and several examples will be given where lattices have been used to represent linguistic concepts.

However, even at this point, a brief historical perspective should hopefully engage those readers who are particularly interested in search engines. As described in Chapter 5, two of the fundamental models for search engines are the *vector model* and the *Boolean model*, and in Chapter 7, we described some of the different and complementary properties that Boolean logic and quantum logic provide when searching and exploring document collections. The underlying geometric model for Boolean logic is set theory, and the underlying geometric model for quantum logic is the theory of vector spaces — and we have just shown that both of these models can be described as lattices.

The difference between Boolean and vector lattices (characterized particularly by the presence or absence of the *distributive* property, page 265) were used by Birkhoff and von Neumann (1936) to describe the logical assumptions of different physical theories. Back in the 1930s, they were not seeking to model the differences between Boolean and vector search engines, but the differences between classical and quantum physics.

This pioneering work firmly rooted the notion that classical physics was based on the logic of set-theoretic lattices and that quantum physics was based on the logic of vector lattices. Understanding the full ramifications of this distinction is one of the key challenges in finding a unified theory that describes both classical physics (including relativity) and quantum theory, which is probably the greatest problem left unsolved by 20th century physics. Now we see that same distinction made manifest in the models we choose to build search engines and navigate concepts in human language: and from a mathematical point of view, the distinction is encapsulated in the fact that Boolean logic and vector spaces are both lattices, but they are lattices with different properties.

> ```
> 1010101
> 0101010
> 1010101
> ``` John von Neumann
>
> John von Neumann (1903–1957) was born in Hungary and worked in Germany and the Uniter States. His doctoral thesis was on set theory, and at the age of 20 he gave the definition for ordinal numbers that is still used today. He contributed to the foundations of game theory, statistical mechanics, continuous groups and representation theory, and he transformed quantum mechanics by developing algebras of operators. In developing the architecture of computer systems he laid the foundations of modern computing, using electronic methods for many applied problems during the second world war.
>
> With a genius for seeing straight to the axiomatic core of engineering problems, the advantage of von Neumann's approach is summed up in his famous quote:
>
> If people do not believe that mathematics is simple, it is only because they do not realize how complicated life is.

8.3 Concept Lattices — Elements and Features

This section will describe two examples of concept lattices, one ancient and one recent, but which have very similar structure. Examples such as these have led Ganter and Wille (1999) to develop a general theory of concept lattices, the main points of which will be described.

8.3.1 The Four Elements

Modern lattice theory began to flower in the first half of the 20th century, but the seeds were sown in ancient times, in particular in the theory of the four elements and their constituent features. As described in Section 1.3.1, the belief that the world is composed of the four elements *earth, air, fire* and *water*, and that different substances arise through different mixtures of these elements, was developed by Empedocles.

Aristotle championed the theory of the four elements in his treatise *On Generation and Corruption*,[5] which in his view both corrected and completed the work of previous philosophers, including the

 Empedocles

Empedocles (*ca.* 490–430 BC) was born in Sicily, and though his works are lost, what we know about his natural philosophy (from Aristotle and other authors) has a curiously modern ring to it (Pullman, 1998, p. 21). As well as the theory of four elements (which correspond closely to the four modern states of matter), Empedocles postulated that all of nature is governed by two kinds of force: *love* (*philia*) which brings things together and *strife* (*neikos*) which drives them apart. In modern science we still divide forces into those of *attraction* and *repulsion*: before presuming that this is much more logical and sophisticated, remember that we too use the term *attraction* to mean something very human and often irrational!

theory of Empedocles and Plato's account of creation given in the *Timaeus*. Discounting Plato's association between the elements and the regular solids (page 21), Aristotle describes the four elements as being combinations of fundamental *qualities* or *attributes*. The basic attributes, in Aristotle's account, are *hot, cold, wet* and *dry*. Each element arises through a conjunction of two of these attributes. If any combination was permitted, this would result in six possible pairings, but the attributes *hot* and *cold* cannot combine because they are opposed to one another, and neither can the attributes *wet* and *dry*. The remaining four combinations of attributes give rise to the four elements, as follows:

---

[5]Though Aristotle elsewhere concludes that "the view that they (the elements) are more than three in number would seem to be untenable." (*Physics*, Bk. I Ch. 5)

Element	Attributes
fire	hot dry
air	hot wet
earth	cold dry
water	cold wet

Another way to visualize these relationships is to draw a *cross-table* (Table 8.1), with the elements down the side and the attributes along the top, and a mark in the corresponding box if a particular element has a particular attribute.

This structure enables us to answer questions such as

Q. What substance is both *cold* and *wet*? A. *Water*.

Q. What attribute is shared by both *fire* and *air*? A. They're both *hot*.

Of course, in many climates, the air is not hot, or the earth is not cold, so this very simple model may not correspond to either physical reality, or the normal meanings of the words in question. However, the mathematical structure of Aristotle's model, and the way he explores it, are very insightful. The opposition between attributes gives rise to opposition between elements: the attributes of *fire* are opposite to those of *water*, and those of *earth* are opposite to those of *air*. The structure that arises is shown in Figure 8.11 (many more artistic representations of this same structure can be found on the internet).

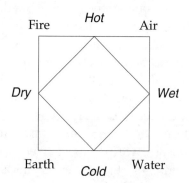

FIGURE 8.11 The opposing qualities of pairs of elements

TABLE 8.1 Cross-table describing the four elements and their attributes

	hot	cold	wet	dry
fire	×			×
earth		×		×
water		×	×	
air	×		×	

This square of elements and attributes is a bit like a doubled-up version of the traditional square of opposition in logic (shown in Figure 8.6), and (as is typical across his many varied works) Aristotle uses much of the same terminology of *contraries* and *contrarieties* to make the analogy with his logical work as clear as possible.

In the 19th and 20th centuries, convincing demonstrations that matter is made up of tiny particles have been taken as strong evidence for the *atomic* theory of Leucippus and Democritus, and the elemental theory of Empedocles and Aristotle has been comparatively discredited. The scientific discoveries that air is a mixture of different gases, and water a compound of hydrogen and oxygen, caused a monumental shift in scientific attitudes. Nowadays, when we hear the term *elements* in science, we are most likely to think of the basic chemical elements (hydrogen, helium, lithium, beryllium, etc.).

A more sympathetic correspondence can be drawn between the four ancient elements and the four modern *states of matter*, which are solids, liquids, gases and plasmas.[6] The description of processes such as evaporation given by ancient philosophers (the adding of heat to water brings about moist air) are not dissimilar to those given by the modern kinetic theory of matter.

 Feature structures in language

One of the main applications of concept lattices in linguistics is to assist in the construction of multilingual dictionaries and translation resources (Janssen, 2002). The traditional approach of preparing a bilingual dictionary for each pair of languages X and Y is impractical, and the use of a concept lattice as an interlingua can potentially widen this bottleneck (and suggest a descriptive translation where there is no direct translation, e.g. by describing *toes* as *fingers of the feet* for translation into the Spanish *dedos del pie*).

Other uses of lattices in linguistics include the *feature structures* of unification based grammars such as LFG and HPSG: a sophisticated account of typed feature structures is given by Carpenter (1992).

For our purposes, Aristotle's model of the elements is of interest not

---

[6]Matter in the *plasma* state consists of particles that are so energetic that the electrons fly away from the nuclei in a burning soup of subatomic particles. Plasmas are manufactured in particle accelerators: our nearest natural concentration of plasma is of course the sun.

TABLE 8.2 Cross-table describing different kinds of *horses* and their attributes

	horse	male	female	adult	young
horse	×				
stallion	×	×		×	
mare	×		×	×	
foal	×				×
filly	×		×		×
colt	×	×			×

because of its scientific accuracy, but because of its descriptive structure. Each element has its particular characteristics because of the presence of certain attributes or features. As we shall see, this makes the concepts and their relationships into a lattice.

8.3.2 A lattice of horses

A modern example of the same theoretical approach has been used to represent the structure of words (in several European languages) that refer to kinds of *horse* (Janssen, 2002, Ch. 2). To describe the differences between the words *horse, stallion, mare, foal, colt* and *filly*, Janssen draws the cross-table shown in Table 8.2. This is very similar to the cross-table for the elements, and if we restrict it to the objects *stallion, mare, filly, colt* and the attributes *male, female, adult* and *young*, the cross-table for the horses and the cross-table for the elements become exactly the same but with different names.

As with the elements, we can use this structure to answer questions about horses (which can enable the choice of a correct translation between languages, one of the reasons for using lattices to build a multilingual lexical database). For example, suppose we wish to know what to call a *horse* that is both *female* and a *foal*. We see that such a creature will have to possess the *young* attribute (to be a *foal*) as well as the *female* attribute, and another look at the table shows that a horse with both the attributes of being *young* and being *female* is called a *filly*.

A structure arising from a cross-table of relations in this way is called a *Formal Context*, and their study is called *Formal Concept Analysis* (Ganter and Wille, 1999). We shall see that a Formal Context gives rise to a natural ordering relationship with meet and join operations — a *concept lattice*.

8.3.3 Intents and Extents

It is time to give formal names to our sets of objects (or elements) on the one hand and to sets of attributes (or qualities) on the other.

Definition 21 Every concept A has an *extent* and an *intent*.
The extent (or extension) of A, written $\mathrm{Ext}(a)$, is the set of objects referred to by A.
The intent (or intension) of A, written $\mathrm{Int}(a)$, is the set of attributes possessed by A.

The reader should be forewarned by the first line of this definition: a very sweeping statement about *all* concepts! Its purpose is to alert the reader to the artificial stipulations being introduced in order to build a mathematical model. This danger is well-recognized by the founders of Formal Concept Analysis, who write

> The adjective "formal" is meant to emphasize that we are dealing with mathematical notions, which only reflect some aspects of the meaning of *context* and *concept* in standard language. (Ganter and Wille, 1999, §1)

This health warning notwithstanding, the very symmetrical relationship of *duality* between intents and extents leads to powerful consequences, enabling us to build whole lattices from the cross-table diagrams in Tables 8.1 and 8.2.

The idea of an *extent* was introduced on page 16 when we were discussing the relationship between concepts, sets, implication and containment. The dual idea of an *intent* is to encode not the set of objects, but the set of features which those objects have in common. A closely related (older) description is to say that the intent of a concept is the set of attributes which are *predicated* on that concept.

8.3.4 Specific examples

Many of the definitions and terms used by naturalists have deliberate intents and extents. For example, the concept **vertebrate** has as its intent the attribute **has_backbone**, and as its extent, the set of all **vertebrates**. In Aristotelean terms, one would say that the attribute **has_backbone** is predicated of all **vertebrates**.

Casting definitions in these fixed terms has enabled naturalists to agree on the classification of previously unknown species, even if only one example of the new species is available. However, it is sometimes counterintuitive — a *whale* is more similar to a prototypical *fish* than to a prototypical *mammal*, and *whales* have often been described as *great fish* (as in many older translations of the Bible story *Jonah and the Whale*, in which the word *whale* is not mentioned).

The *killer whale* has the appearance of a whale but the intent of a dolphin.

In another example involving marine mammals, the *killer whale* is actually the largest member of the *dolphin* family, but because of its size, it is commonly called a whale.[7]

The problem of distinguishing the attributes that are central to the classification of a species from the attributes that are incidental is humourously raised in *Alice in Wonderland*:

> "I *have* tasted eggs, certainly," said Alice, who was a very truthful child; "but Little girls eat eggs quite as much as serpents do, you know."
>
> "I don't believe it," said the Pigeon; "but if they do, why, then they're a kind of serpent: that's all I can say." (Carroll, 1865, Ch. 5)

8.3.5 Meets, Joins and Ordering

There is a natural ordering relationship on concept lattices which is quite familiar to us. As normal, we regard the concept B as a hypernym of the concept A if the extent of A is contained in the extent of B (i.e. $A \sqsubseteq B$ if and only if $\text{Ext}(A) \subseteq \text{Ext}(B)$). Using the extents of concepts therefore gives a simple ordering relationship to a concept lattice, which corresponds to the important hypernym relationship $\sqsubseteq$.

There is a mirror-image or *dual* version of this relationship, which involves reasoning with intents rather than extents. If $A \sqsubseteq B$, any object described by A will have to possess at least all of the attributes possessed by objects described by B. It follows that the intent of A *contains* the intent of B, or $\text{Int}(B) \subseteq \text{Int}(A)$.

---

[7]In fact, the name *whale killer* might be a more precise name, since killer 'whales' actually prey on other whales. Swapping the words round makes a tremendous difference!

This dual relationship can be summarized by the equivalences

$$\text{Int}(B) \subseteq \text{Int}(A) \iff \text{Ext}(A) \subseteq \text{Ext}(B) \iff A \sqsubseteq B. \qquad (8.22)$$

Concepts in a concept lattice are ordered identically by either increasing *extent* or decreasing *intent*, and these are the same. To get an intuitive feel for Equation 8.22, think of the extent as 'people eligible to vote' and the intent as 'conditions voters must satisfy.' The stricter the conditions or attributes required for eligibility (traditional conditions have included being *male, adult, propertied, a citizen*), the fewer potential voters there are who possess all of those attributes: on the other hand, if there are *fewer* conditions of eligibility, more people are potentially eligible to vote.

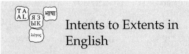

Intents to Extents in English

In English, attributes are often described using adjectives. For example, the adjectives *rich*, *hungry* and *living* can all be used as attributes which modify the description of a person.

These descriptive words can all be prefixed with the word *the*, which coerces the adjective into behaving as a noun referring to a class of people — for example, *the rich* means the set of rich people, *the hungry* means the set of all hungry people, and *the living* means the set of all living people.

In these cases, the same word can be used both to describe an intensional *attribute* of a concept and an extensional *set of objects* which have that attribute.

We can now go about the process of forming meets and joins of concepts, bringing our analysis of simple cross-tables into lattice theory. These can be framed in terms of both intents and extents. For example, consider the concepts *colt* and *filly*. Their least upper bound or join is given by the concept *foal*, since the extent of *foal* is the smallest set that contains the extents of both *colt* and *filly*. Using set theoretic symbols, we could write

$$\text{Ext}(colt) \cup \text{Ext}(filly) = \text{Ext}(foal). \qquad (8.23)$$

There is a corresponding or *dual* version of this reasoning. The intents of *colt* and *filly* are given by

$$\text{Int}(colt) = \{young, male, horse\}$$

and

$$\text{Int}(filly) = \{young, female, horse\}.$$

For an object to be either a *colt* or a *filly*, it must at least have the attributes {*young, horse*}, because these attributes are common to both *colts* and *fillies*. These attributes are precisely the ones that define the concept *foal*. This reasoning can be expressed in the dual form of Equation (8.23),

$$\text{Int}(colt) \cap \text{Int}(filly) = \text{Int}(foal). \tag{8.24}$$

We see that by taking the *union* of the extents, or the *intersection* of the intents, we arrive at the same concept. This complementary or *dual* point of view arises form the natural ordering of objects in our Formal Context.

The Hasse diagrams of the concept lattices for the four elements (Table 8.1) and for the different horses (Table 8.2) are shown in Figure 8.12. These diagrams should be interpreted as follows:

- Each node corresponds to a concept
- The extent of each concept is written underneath the node
- The intent of each concept is written above the node
- The extent of each concept is contained in the extent of every concept *above* it in the graph
- The intent of each concept is contained in the intent of every concept *beneath* it in the graph

If you remember that extents propagate upwards and intents propagate downwards, you've grasped one of the most important points. In the lattice of horse names, many of the labels have been left off, leaving only labels on the *lowest* instance of each object, and only on the *highest* instance of each attribute. The remaining labels can be easily filled in by propagating the extents upwards and the intents downwards (a useful exercise).

To be a lattice, every pair of concepts must have a meet and a join. This motivates the insertion of nodes for concepts that do not have individual names attached to them. In particular, at the top of every concept lattice, there is a node representing 'every object,' which may not have a name. For example, the top node in the four-element lattice is labelled *earth, air, fire, water*, and it is (for example) the join of the opposing concepts *fire* and *water*. If this node were to have a name attached to it, we might choose a very general term like *matter* or *substance*.

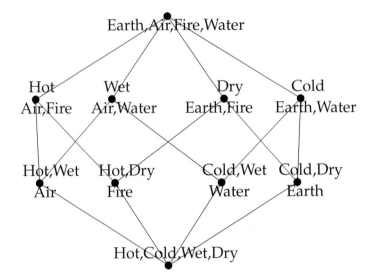
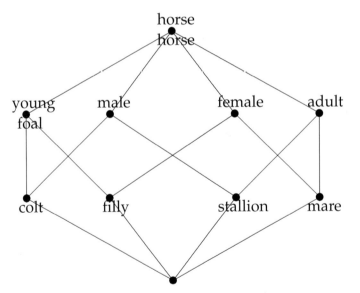

FIGURE 8.12 Concept lattices of the four elements and words for horses. The lower diagram shows only the essential labels, though the others can be added by propagating extents (written below their nodes) upwards and intents (written above their nodes) downwards. Drawings created by JaLaBA, Maarten Janssen (2001).

In terms of intents, the top node is devoid of attributes and the bottom node has *every* attribute. If any of the attributes are incompatible, this bottom node will have zero extent. (For example, in the four-element lattice, there is no substance that is *hot* and *cold* and *wet* and *dry*.) Lattices of sets always have a bottom node given by the empty set; similarly, vector-space lattices also have a unique bottom node because the zero point or origin is contained in every subspace. The top and bottom nodes also correspond to the truth values 1 and 0, a convention introduced by Boole, who wrote, right at the beginning of his syllabus:

> Let us employ the symbol 1, or unity, to represent the Universe, and let us understand it as comprehending every conceivable class of objects.

(Boole, 1847, Ch. 1)

Ontology management using Formal Concept Analysis

Integrating several knowledge bases or ontologies has become an important goal for information management, since developing a new knowledge base for every new application is costly, wasteful, pretty unimaginative, and certainly not scalable enough to make the Semantic Web a reality.

The lattice based methods of Formal Concept Analysis have been used to address this problem, by regarding concepts in each prior ontology as intents, and documents where those concepts are referred to explicitly as extents. This enables two ontologies K_1 and K_2 to supply intensional information relative to a common extensional document collection D. Suitable concepts in K_1 and K_2 are then merged if their extents in the document collection are similar, giving a bottom-up algorithm called FCA-MERGE (Stumme and Maedche, 2001).

The root node in a hierarchy (page 86) also corresponds to this universal concept. A hierarchy or taxonomy of objects can be though of as 'half of a concept lattice': the half that deals with sets of objects and extents, omitting the dual half that deals with sets of attributes and intents.

The top node of a concept lattice does not need to represent the whole of Boole's universal set — many lattices are more specialized than this. For example, if we restrict our attention to the domain of horses, the top node is just the general concept *horse*, of which the other nodes are subconcepts. Similarly, in the lattice of factors of the number 60

(Figure 8.7), any other number n could be used as the top node, giving a lattice which is special to the number n.

One of the benefits of examining a collection of smaller lattices is that many attributes are relevant only to certain branches of a concept lattice. For example, in our lattice of horses, every member is either *male* or *female*, but in our lattice of numbers, this distinction makes no sense (though not surprisingly, in some cases genders *have* been assigned to numbers and musical notes; see Daniélou 1943, Ch 2). Aristotle notes this point, saying:

> If genera are different and coordinate, their differentiae are themselves different in kind ... But where one genus is subordinate to another ... the greater class is predicated of the lesser, so that all the differentiae of the predicate will be differentiae also of the subject.   (*Categories*, Bk, I Ch. 3)

This is strikingly similar to some of the ideas we have been using in concept lattices. If we explained that words like *differentiae*, and *attributes* were being used to mean similar things, and that the collection of attributes belonging to a concept was called its intent, Aristotle would probably be fully conversant with the rule that 'intents propagate downwards.'

This introduction to lattice theory, particularly concept lattices, has been extremely brief, and the expert will see that many important points have been left out. The interested reader should turn to Davey and Priestley (1990), Ganter and Wille (1999), and the classic text of Birkhoff (1967) for more information.

8.3.6 Summary of concept lattices

A lattice is therefore a set with an ordering *relationship* $\preceq$ and two *operators*, join (written $\vee$) and meet (written $\wedge$).

In concept lattices, the ordering of concepts is the same as the ordering relationship $\sqsubseteq$ in concept hierarchies. The meet and join operations correspond to conjunction and disjunction of concepts. This logical structure should be apparent in the concept lattices of Figure 8.12 — for clarification, see the picture in Figure 8.13 showing conjunctions and disjunctions of concepts in the horse-lattice.

Fortunately, the notation that has become standardized in lattice theory was deliberately chosen to help clarify the link between logic and lattices. The notation $a \vee b$ is used to express the disjunction in logic and

Concept Lattices: Binding Everything Loosely Together / 279

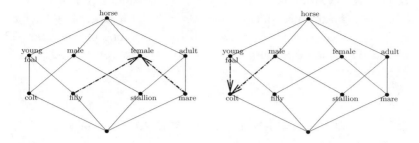

FIGURE 8.13 The disjunction of *filly* and *mare* and the conjunction of *young* and *male* in the horse-lattice.

the join in lattice theory, and these concepts are equivalent. The notation $a \wedge b$ is used to express the conjunction in logic and the meet in lattice theory, and these concepts are equivalent.

8.4 The Axioms of Logic — Lattices with Negation

Once we have models for implication, conjunction and disjunction, the final piece of apparatus needed to turn a lattice into a *logic* (in the mathematical sense) is a *negation* operator. Traditionally, the negation operator is different from conjunction and disjunction because it is a *unary* rather than a *binary* operator, meaning that it takes only one input rather than two. (See Figure 7.1 on page 206 for a refresher.)

To add a suitable model for negation in a lattice, we introduce the notion of a *complement*.

Definition 22 Two nodes a and b are called *complements* of one another if $a \vee b = 1$ and $a \wedge b = 0$.

For example, *male* and *female* are complements of one another in the horse-lattice — their join is the top node (called 1) and their meet is the bottom node (called 0). The definition of complements is designed, to some extent, to reflect the properties embodied by *opposites* in natural language, though as we shall see, many words in everyday language don't really have an opposite of this sort, and this is one of many important examples which demonstrates that language does not conform to the simplifications of formal logic.

The idea of finding complements is partly motivated by

considerations from Boolean logic. If $\neg a$ is the complement of a, $a \wedge b$ stands for the meet of two elements and $a \vee b$ for their join, 1 stands for *True* and 0 stands for *False*, then the conditions in Definition 22 become the standard *complement laws* of logic (Partee et al., 1993, p. 110):

$$a \vee (\neg a) \text{ is true.} \qquad a \wedge (\neg a) \text{ is false.}$$

or less formally

$$a \textbf{ OR } (\textbf{\textit{NOT}} a) \text{ is true.} \qquad a \textbf{ AND } (\textbf{\textit{NOT}} a) \text{ is false.}$$

Note also that Definition 22 implies that the top element 1 and the bottom element 0 are complements of one another — so *True* and *False* are always opposites.

Under the right conditions, a lattice where every element has a complement in this way is called a *logic*. Axioms for a general logic are given in Figure 8.14. However, many authors (including Hamilton (1982) and Davey and Priestley (1990)) only give the corresponding definition for a *Boolean* logic (sometimes called a Boolean algebra), which is actually *simpler* because the distributive law takes care of some of the other axioms for you. While I have tried to present the features of a *general* logic, the fact that I had to turn to the work of Varadarajan (1985) on quantum geometry to find the axioms in Figure 8.14 written down explicitly suggests that, for the most part, non-Boolean logics remain a rather specialist field. This is a pity, because non-Boolean and particularly non-distributive logics occur very naturally, and are appropriate formal models in situations which involve smoothing and reasoning by induction, including many machine learning applications and spatial databases (see Widdows and Higgins, 2004).

In fact, neither the elements-lattice nor the horse-lattice of Figure 8.12 are logics in this sense. To see this, note that complements are not uniquely defined in these lattices. For example, the *filly* node is a complement of *stallion*, *male* and *adult*! In order to be a logic, each element in a lattice must have a *unique* complement — otherwise the step from complementation (a 2-place *relationship*) to negation (a 1-place *operator* or mapping) is not well-defined. On the other hand, if every node a in a lattice L has a unique, well-defined complement which satisfies the axioms in Figure 8.14, then this complement is considered to be the *negation* of a, often denoted by the symbol a' in lattice theory and $\neg a$ or $\sim a$ in logic.

> **Formal Definition of a Logic**
>
> A *logic* is a lattice L together with an operation $\neg : L \to L$ such that
>
> - N1. $\neg$ is a bijection.
> - N2. $a \preceq b$ implies that $\neg b \preceq \neg a$. (So $\neg$ is an *order-reversing map*.)
> - N3. $\neg(\neg a) = a$ for all $a \in L$.
> - N4. $a \wedge \neg a = 0$ for all $a \in L$.
> - N5. $a \vee \neg a = 1$ for all $a \in L$.
>
> In addition, the lattice L must satisfy
>
> - For any countably infinite sequence $a_1, a_2, \ldots$ of elements of L,
> $\bigvee_j a_j$ and $\bigwedge_j a_j$ exist in L.
> - If $a_1 \preceq a_2$, there exists some $b \in L$ such that $b \preceq \neg a_1$ and $b \wedge a_1 = a_2$.

FIGURE 8.14 A logic must satisfy these axioms (Varadarajan, 1985, p. 42).

It is worth noting that the last condition in Figure 8.14 is related to a note of Aristotle's, namely:

> If A is attributable to no B, then either this predication will be primary, or there will be an intermediate term prior to B to which A is not attributable.
> (*Posterior Analytics*, Bk I, Ch. 19)

8.4.1 Orthogonal complements and quantum logic

A good example of a complement operation which turns a lattice into a logic is given by the *orthogonal complement* in vector spaces. We have already described the collection of subspaces of a vector space in some detail, as a geometric structure, as a lattice, and as the spatial model for quantum logic. To round off this picture, this section highlights the role of orthogonality in making the Grassmannian vector space lattice described in Section 8.2.4 into the quantum logic of Birkhoff and von Neumann (1936), described in Section 7.6.

Let U be a subspace of a vector space V, and let $U^\perp$ be the *orthogonal subspace* to U. In other words, $U^\perp$ is the collection of vectors which are orthogonal to every element of U. For example, in the 3-dimensional geometry of Figure 7.4 on page 228, the vector C is orthogonal to every vector in the plane AB. This relationship can be written down using the symbols $(AB)^\perp = C$, or conversely $C^\perp = AB$.

It is easy to see in this example that $AB \cap C = 0$ (the line C and the plane AB only intersect at the zero point) and that $AB + C = \mathbb{R}^3$ (you can get to any point in 3-dimensional space by travelling to a given point in the plane AB to get to the desired horizontal coordinates, and then a given vertical distance along the line C). This pair of subspaces therefore satisfies Definition 22, so the plane AB and its orthogonal subspace C are complements of one another.

More generally, all pairs consisting of a subspace U and its orthogonal space $U^\perp$ have the following interesting properties:

- For any subspace U of a vector space V, its orthogonal complement $U^\perp$ is also a subspace of V.
- Taking the orthogonal complement twice gets you back to the space you started with, or in symbols $(U^\perp)^\perp = U$.
- The intersection of U and $U^\perp$ is just the zero vector 0, or in symbols $U \cap U^\perp = 0$.
- The vector sum of U and $U^\perp$ is the total space V, or in symbols $U + U^\perp = V$.

The first of these (very naturally) ensures that if a concept is represented by a subspace, then the orthogonal space is also a well-formed concept — every concept must have a well-defined negation in logic. The second point corresponds to the double-negative rule

$$NOT\,(NOT\,p) = p.$$

The third and fourth observations show that U and $U^\perp$ are always complements of one another. To emphasize this correspondence, the orthogonal subspace $U^\perp$ is often called the *orthogonal complement* of U, making this relationship explicit.

As an easy exercise, the reader may match up these four observations to four of the five axioms of logic in Figure 8.14. That the rest of the axioms of logic hold in this situation can also be demonstrated without difficulty — orthogonal complementation has precisely the right properties to turn the lattice of subspaces of a vector space into a *logic* of subspaces. Just as the set complement (page 14) is the model for negation in Boolean logic, so the orthogonal complement is the model for negation in the logic of vector spaces, known as quantum logic.

It is interesting to note that the presentation of quantum logic given by Birkhoff and von Neumann (1936) introduces the Aristotelean apparatus first, basing the axioms of quantum logic initially upon implication and negation, to which the set intersection ('set product') and vector sum ('closed linear sum') can be added to give a logic. It is only subsequently that these authors approach quantum logic through the axiomatic development of ordered sets, lattices, and finally complementation. These sections (Birkhoff and von Neumann, 1936, §6–9) remain one of the clearest expositions of the development of logic from an axiomatic point of view, the story I have tried to outline in this chapter.

The notion of complementation is very important in Grassmann's geometric algebra, beginning with the following definition:

> If E is a unit of any arbitrary order (i.e. either one of the primitive units or a product of a number of them) the *complement* of E is defined as a quantity E' of all units which are not in E. (Grassmann, 1862, §89)

In other words, in a system where every concept can be expressed as a combination of primitive units, the complement of any concept E is the concept E' formed by combining all the *other* units. Thus E and E' have no factors in common, but between them, E and E' contain *all* the primitive factors.

In a vector space of dimension n, the subspaces of dimension k and the subspaces of dimension $n - k$ are complements of one another, in both quantum logic and in the Grassmannian geometry from which quantum logic stems. For example, in three dimensions, the set of all lines through the origin and the set of all planes through the origin come in complementary pairs, the complement of each plane being the line orthogonal to it, and the complement of each line being the plane orthogonal to it. This is often implemented by using the scalar product to determine which lines and planes are orthogonal to one another, though Grassmann originally did this the other way round, using his symbolic algebra to define complementation and *then* using the idea of complementation to define a general inner product (Grassmann, 1862, §137), of which the scalar product is one specific example.

8.4.2 Boolean logics and lattices of sets

There are many other modern logics, some of the most important being the *Boolean* logics, each of which is equivalent to some lattice of sets (Partee et al., 1993, §12.3). One of the simplest is shown in Figure 8.15, which shows the lattice of subsets of the three-element set $\{a, b, c\}$.

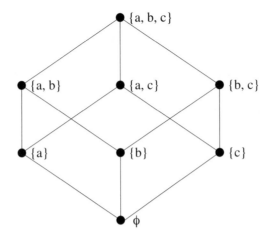

FIGURE 8.15 The Boolean logic of subsets of the set $\{a, b, c\}$.

(The symbol ϕ is used to represent the empty set.) The usual ordering (containment), meet (intersection) and join (union) operations can easily be read from this diagram.

The lattice can be thought of as four pairs of opposites, as follows:

$\{a\}$ is opposite to $\{b, c\}$
$\{b\}$ is opposite to $\{a, c\}$
$\{c\}$ is opposite to $\{a, b\}$
$\{a, b, c\}$ is opposite to ϕ

Each of these pairs are complements of one another in the sense of Definition 22, and the mapping from a subset to its complement is a well-defined negation operator on the lattice of sets, which makes the lattice into a logic.

Such a diagram could be drawn for the subsets of a set U of any size (though the size of the diagram would double every time a single element is added to U). All such diagrams can be drawn showing the same elegant symmetries, in which a rotation of 180 degrees maps each element onto its unique complement.

8.4.3 Combined states in lattices

A Boolean logic (or lattice of sets) is a very special kind of lattice, where every possible combination of elements is a subset or concept in its own right. This is often not the case in language — for example, we have no word in English for the set consisting of *dogs, fish* and *spiders*.

This shortfall — that many possible sets do *not* form the extent of any one concept — is one factor that makes many concept lattices much more closely related to quantum logic than to Boolean logic. This is because finding the join of two concepts involves more than just taking the set union of their extents — it may also involve the inclusion of combined states, as other concepts are subsumed by the outcome of the join operation. This process can sometimes be modelled geometrically using *closure conditions* — for example, if stable concepts are represented by *convex* sets (an approach proposed by cognitive scientists including Gärdenfors (2000)), the disjunction of two concepts will include any combined states necessary to ensure that the resulting concept also is represented by a convex region (Widdows and Higgins, 2004).

For example, in the horse-lattice, the join of *stallion* and *filly* is the top node *horse*, which

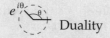

Duality

Words like *duality* and *dualism* are often used to describe a conflict between opposing forces (such as *good* and *evil*) or descriptions (such as *mental* and *physical*). In mathematics, *duality* is used more gently to describe different approaches which are equivalent and complementary — duality holds up a mirror, and sometimes problems are easier to solve on one side of the mirror, sometimes on the other.

Lattice theory thrives on duality. All of our lattices can be drawn the other way up, and would still contain the same information, which is why so many of them have been naturally symmetrical. A good metaphor for duality in lattices is to think of Figure 8.15 as a picture of the famous *Necker cube*. Is the cube going into or coming out of the paper? Neither interpretation is wrong, and both are insightful.

Duality is particularly powerful in projective geometry, where statements and theorems often come in pairs or *duals*, such as "2 lines join in a plane" and "2 planes meet in a line." With such pairs, we can often prove one statement and get the other for free.

subsumes all of the other nodes including *mare*. It follows that *mare* is contained in the disjunction of *stallion* and *filly*, or in symbols,

$$\textit{mare} \sqsubseteq (\textit{stallion} \vee \textit{filly}).$$

Interpreted in a Boolean fashion, this would mean that every *mare* is either a *stallion* or a *filly*, which is nonsense. Instead, the correct interpretation is that the attributes of *stallion* (in particular, being *adult*) and the attributes of *filly* (in particular, being *female*) can between them give the attributes of *mare*.

Another feature of many concepts in language is that there is no name given to their complements, even if such complements exist. For example, there is no word in English for *not-horse* which can be predicated of all objects which are not horses. Even within the small horse-lattice, it is true to say that *colts* are not *stallions*, *mares* are not *stallions* and *fillies* are not *stallions*, but there is no concept with the extension {*colts, fillies, mares*} which can be considered as the unique complement of the concept *stallion*.

Even though negation is regarded as a vital ingredient of modern logic, this lack of words for uniquely defined complements or opposites should not surprise us. A negated concept such as *not-horse* is very uninformative — nearly every object in the universe is a *not-horse*, and we need to express this concept so rarely that having a special word for it would be silly. Natural language arose for the sake of conveying information, not to satisfy the axioms of logic!

Aristotle faced this question diligently. Starting with substances, he posed and answered the following question:

> What can be the contrary of any primary substance, such as the individual man or animal? It has none. (*Categories*, Ch. 5)

The other categories are similarly examined — relatives, actions, affections, and some qualities have contraries, whereas species, genera, and quantities have none. Though we may not agree with these distinctions, we must certainly respect the thought behind them. It should be noted that lexicographers *have* followed Aristotle's model with more diligence — for example, WordNet clearly records which words have antonyms (in particular, most adjectives) and which words do not.

Conclusion

This chapter has shown the way in which many of our models — particularly hierarchies, vector spaces, and logics, can be regarded as kinds of lattices, a unifying concept whose full ramifications have yet to be explored.

There are still some aspects of logic (such as quantification) and many aspects of language which we have not brought into the framework of lattice theory — not in this chapter, at least. Some final remarks about these challenges are made in the final conclusion to the book.

Wider Reading

One of the greatest works on lattice theory remains Garrett Birkhoff's comprehensive textbook (Birkhoff, 1967), which contains a thorough analysis of the distributive law and the ensuing differences between classical and quantum logic — and many chapters on lattices that appear in other parts of mathematics such as topology and group theory, which will be of interest to mathematicians. One of the best summaries of the way logic and lattices are intertwined is still the paper of Birkhoff and von Neumann (1936), which remains one of the landmarks of the field.

A more recent and accessible standard text on lattice theory is that of Davey and Priestley (1990), which describes partially ordered sets and Hasse diagrams, the theory of Formal Concept Analysis and many applications in computer science, including domains and information systems. Concept lattices themselves are introduced much more fully by Ganter and Wille (1999), and their multilingual applications by Janssen (2002).

In language technology, the most widespread applications of lattice theory can be found in speech recognition (the same can be said of other sophisticated mathematical models). Examples include the papers of Chappelier et al. (1999), Li et al. (2002), and Kacmarcik et al. (2000). The use of lattice-based models such as typed feature structures in natural language processing is discussed by Carpenter (1992), this work being related particularly to unification-based frameworks such as Head-Driven Phrase Structure Grammar (Pollard and Sag, 1994).

The idea of the specific-to-general ordering given by implication is important in machine learning (Mitchell, 1997, Ch 2). The extent to which a logical disjunction should be non-distributive is one of the key factors in determining the appropriate *inductive bias* which an automatic learner should use to generalize from specific training examples.

There are great problems facing any attempt to define linguistic concepts by listing the attributes that those concepts must possess — for example, Wittgenstein (1953, §66) points out that there are a multitude of activities we call *games* and no single feature that all of these activities possess. (Not all games are competitive, not all games have rules, not all games are amusing, but they are all recognizably games.) These problems are sometimes used to motivate complementary ideas of concepts such as those given by prototype theory (Aitchison, 2002, Ch. 4, 5).

The naïve view that concepts *can* be defined by listing their necessary and sufficient features is sometimes described as an 'Aristotelean' approach (see for example Gärdenfors 2000, §3.8). While Aristotle pioneered the development of such a system of classification for biology (while aware of its limitations: see e.g. *Parts of Animals*, Bk 1, Ch. 1), his careful analysis of the variety of meanings available for some of the most important philosophical terms (*Metaphysics*, Bk 5), and the phenomena of ambiguity more generally (*Topics*, Bk 1, Ch. 15), demonstrate no facile presumption that the method should be applied indiscriminately. Aristotle's views *have* however inspired some modern linguists including Pustejovsky (1995) to approach ancient problems such as systematic polysemy using a variety of relationships and mappings involving type-lattices.

Other reading and future directions will be suggested in the final conclusion to the book, which follows.

Conclusions and Curtain Call

Our journey through geometric spaces of words and their meanings draws to a close, and it is time to meet some of the main characters one last time, and to reconsider the progress that has been made in the broader setting of science at the beginning of a new century. The linguistic challenges that still need to be addressed are enormously diverse and elusive, and our theoretical models still only capture a fraction of the intuitive adaptability used by humans in perfectly normal communication. My aim is not to leave the reader falsely confident, but hopeful, curious, and excited at the thought that the some of the most important scientific discoveries still await us, whatever they may be.

A Conceptual Structure of Conceptual Structures

This section reviews the progress we have made and tries to describe all of our models together, insofar as this can suitably be done in a few paragraphs. My goal is that the reader who has made it thus far should leave with a clear summary of the different strengths of geometric and logical models for representing concepts, and the varying properties which account of the differences between these models.

Putting great faith in our own methods, I decided that the best way to attempt this was to build a concept lattice describing the structures themselves. The objects of this lattice are the various geometric models we have investigated, namely

> Graphs, Hierarchies, Vector space, Ordered sets, Boolean logic, Quantum logic, Lattices, Aristotelean logic.

The main properties of this lattice are the relationships and operations

TABLE 8.3 Cross table describing different kinds conceptual structures and their properties

	Implication	Proximity	Quantification	Negation	Disjunction	Conjunction	Represent documents
Ordered set / DAG	×						
Graphs		×					
Hierarchies	×	×			×		
Vector space		×					×
Quantum logic	×	×		×	×	×	×
Concept lattices	×	×			×	×	
Boolean logic	×			×	×	×	×
Aristotelean logic	×		×	×			

we have investigated in this book, which are well-defined in some models but not in others. These include

> Implication, Proximity, Disjunction, Conjunction, Negation, Quantification, and the ability to Represent Documents Automatically.

An approximate cross table for this collection of objects and properties is given in Table 8.3, and the corresponding concept lattice is given in Figure 8.16.

This lattice gives a pretty good summary of the paths that have been covered in this book. Three key properties are having good representations for similarities or distances, having a notion of implication or containment, and having ways to automatically represent documents as well as terms. We have described how these and other properties contribute to the geometric and logical characteristics of several interesting models, and described a variety of techniques which have been used to build these models automatically and by manual labour (or automatically by using the previous manual labour of others).

Figure 8.16 is *not* a good summary of what methods are actually *possible* in different models. For example, quantification is a perfectly natural idea in Boolean and quantum logic, but we have not investigated it in these settings. We discussed the notions of similarity and proximity

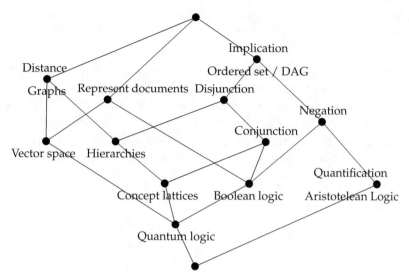

FIGURE 8.16 Concept Lattice of Conceptual Structures. Drawing created by JaLABA, Maarten Janssen 2001.

in graphs and hierarchies in Chapter 4, and a similar discussion of set-overlap measures would also be possible within the framework of Boolean logic. There has been much good work on representing documents in hierarchies which has not been covered in this book (see for example Dumais and Chen 2000, Chakrabarti et al. 1998), though these methods are only partially automatic in that they rely on hand-built training sets.

One might conversely argue that quantum logic (i.e. Grassmannian algebra) is not such a good method for determining implication, because there is so little difference between being *in* a subspace and being *near* a subspace that drawing exact inferences from empirical data is difficult. The built-in notions of distance and similarity given by the Euclidean distance, scalar product, and a range of other measures can be used to 'thicken' low-dimensional subspaces, but this leaves a broader version of the same problem — where do you draw the boundaries?

A model for quantification *is* notably lacking from many practical information systems — many search engines will tell you how many documents match the keywords in your query, but do not tell you which of these matching documents you should read in order to become fully acquainted with the topic in question. (This varies from user to user and situation to situation, making the problem even more complicated.)

Whatever the drawbacks of this description, I hope that Figure 8.16 is useful in summarizing the book so far, and suggesting which areas of future research might be fruitful. In particular, there is currently no model occupying the optimal lowest node — in other words, there is no system which possesses every beneficial property. It is my hope that Figure 8.16 will over the years acquire many new links and creative new objects and properties, as the field (and the author's expertise) progresses.

Last linguistic points

A much more serious objection to the classification of models in Figure 8.16 is the question of whether any of these spaces usefully represents *linguistic* properties. We may be able to say which models have negation or disjunction operators in a mathematical sense — but how well do these operators represent what people mean when they use the words *not* and *or* in natural language?

With these two operators, we have made at least some progress in this book by exploring *many kinds* of negation and disjunction, not just one, demonstrating that the natural negation operators in Boolean and vector search engines exhibit respectively local and non-local behaviour. While our quantitative analysis of these differences has been new, the insight that words like *not* and *or* are used with a variety of meanings is ancient (though often overlooked).[8]

Even with these simple logical connectives, the linguistic situations where they are appropriate are not determined simply by finding uses of the simple words *and, not* and *or*. For example, the title "Romeo and Juliet" does not imply that the play is about some concept given by the intersection of the set {*Romeo*} with the set {*Juliet*}, which would be zero. It is more accurate to say that "Romeo and Juliet" refers to the combined set {*Romeo, Juliet*}, which as we have seen corresponds to the set union or *disjunction* of {*Romeo*} and {*Juliet*}. Most of the links in the graph model of Chapter 2 are learned from corpus examples of just this kind.

---

[8]See in particular Aristotle (*On Interpretation*, Ch 10), Horn (2001), and Gamut (1991).

In some situations, this flexibility between conjunction and disjunction comes naturally with lattice theory, where duality always enables you to look at things both ways. For example, consider the question

What is x if it's an oak, an ash, a fir or an apple?

An appropriate answer would be that x is at least a tree, which is given by the traditional interpretation of the word *or* as a disjunction operation upon the extents of the concepts, *tree* being the most specific concept whose extent includes the extents of *oak, ash, fir* and *apple*.

An alternative (and more normal) way to phrase this question is to ask

What do an oak, an ash, a fir and an apple have in common?

This dual version phrases the question in terms of the attributes or intents of the concepts *oak, ash, fir* and *apple*, and the word *and* can still be interpreted as a conjunction: only it is a conjunction upon intents rather than extents. We rule out those attributes such as *bearing acorns* or *bearing edible fruit* which are not possessed by all of the concepts, which still leaves those attributes which enable us to classify these concepts more broadly as *trees*— which is of course the same answer we obtained using the dual approach on extents, above. Considerations like this are sometimes used to resolve ambiguity — "coal and ash" almost certainly refers to *ash* as a residue of combustion, whereas "oak and ash" almost certainly refers to *ash* as a kind of tree (Resnik, 1999).

As we've seen, ambiguity is an intriguing and enormous challenge whenever one is working with natural language and computers, because ambiguity is naturally exploited by humans but computers are practically allergic to it. Any formal model that claims to represent *the* meaning of a word, in the presumption that there are no other meanings, will sooner or later fall flat on its face. For example, throughout Section 8.3.2, we represented the concept *mare* with the attributes *horse, female, adult,* but this word is sometimes used in English to refer to *all* female horses. The same ambiguity exists for the word *women,* which usually refers to all human females in the phrase "men and women," but only *adult* human females in the phrase "women and children." Often a word is used to refer to creature of a particular gender and species is also used to refer to the whole species: *cows* and *bulls* are all *cows, dogs* and *bitches* are all *dogs, tigers* and *tigresses* are all *tigers*. (Indeed, the choice of whether a female or male term is used for the whole species sometimes reveals some embarrassing bias based on whether the animal is regarded as active or passive.)

This variety among meanings is amplified further by the simple but subtle ways in which people combine words to generate new meanings, something which becomes very apparent when considering examples in English where words are juxtaposed (Cruse, 2000, Ch 4). There are a few examples of 'Boolean conjunction' where the combination may be regarded as the intersection of two sets — for example, *white sheep* refers to the intersection of the class of *white objects* with the class of *sheep*. But *white wine* is only *white* relative to *red wine*, and of course healthy people's skin is never "white as snow." Once we talk about *White Russia*, the colour has come to mean a political affiliation, and by the time we have a phrases like

> Asked recently about his relationship with the White House, Moynihan responded: "There's an advise and consent sort of relationship."
> (New York Times, July 1994)

the *white* refers to the colour of the walls of a building which is being used metaphorically to represent a political institution.

These examples are no problem to human readers, and when the array of different meanings of *white* is drawn to people's attention it causes more amusement than difficulty. This only makes it harder to design good models and computer programs which can demonstrate the same effortless flexibility and robustness.

The empirical approaches we have described in this book are often a big help with these problems. For example, one can compare the distributions of phrases such as *cox apple* and *Big Apple* and work out that the former is likely to be a kind of *apple*, but the latter is not (Baldwin et al., 2003). This approach relies on our ability to combine geometric models such as WORDSPACE with a certain amount of grammatical processing before we start, in order to recognize phrases which should be treated as whole units rather than as separate parts. In the other direction, we have already seen that there are automatic ways of deducing that old words are being used with new meanings, and to try to give labels to those new meanings (see Sections 4.6, 6.3 and 6.7).

Different models are more or less suitable for representing different aspects of meaning, and one size rarely fits all. For example, Boolean logic has great problems with assigning meanings to compounds such as *tiger moth* and *tiger economy*, neither of which is a *tiger*. In many compounds of this sort in English, the first word is being used not as a literal categorization, but as a modification of some feature of the final or *head* word, which usually tells us what kind of object we're talking about.[9]

Concept lattices can solve the problem of the *tiger moth* much more elegantly, because they keep track of both the intents and the extents of concepts. Trying to combine the extents of *tiger* and of *moth*, we realize that their set intersection is empty because nothing can be both a *tiger* and a *moth*. So, suppose instead that it is some attribute from the *intent* of *tiger* which is contributing to the meaning of *tiger moth*. Weighing hundreds of pounds, being (by reputation) fierce, and having a backbone are not very likely features for a *moth* to possess, but being striped like a tiger is very reasonable: the representation of *tiger moth* in the lattice should therefore be like the representation of *moth* but with an extra attribute added, that of being **striped**. Similarly, a *tiger economy* certainly can't be **striped**, but it might be **fierce**, this being the correct interpretation. Such an outline has also been suggested in cognitive science (Gärdenfors, 2000, §4.4): some version of this approach will almost inevitably be necessary, because humans so often name things not only because of what they *are*, but also because of what they *resemble*.

There are many many more examples of this kind, compatible and confusing for different theories and models, and some which on the face of it seem impossible to model at all without squaring the circle. One message of this section is that, in spite of all of our progress, language is always liquid enough to slip through our formal fingers. But the other message is to be hopeful and keep trying things out: there *are* good models which are appropriate for dealing with many different features of language very appropriately in certain contexts, and there has already been tremendous progress.

---

[9]In many other languages it's the other way round. For example, in traditional Latin terms for species of plants and animals, the head word comes first, e.g. the common *rock pigeon* is officially the *columba livia*, and it is the word *columba* which tells us that this species is a member of the pigeon family.

Last mathematical points

Given its unbounded variation and productivity, what aspects of mathematics are most fitted to help truly understand and model what is going on in human language? In this book we have used graph theory, linear algebra, lattice theory and logic; statistical and probabilistic models have been used with great success as well (Manning and Schütze, 1999). The distinction between these various methods is often in the eye of the beholder: our graphs and vectors are learned statistically, and quantum logic in vector spaces is one of the most thorough probabilistic calculi yet developed (Wilce, 2003).

Since these methods have so many overlapping features, we are left to ask which of these features are best fitted for representing language. Questions of this sort are traditionally fertile in mathematics, because they force us to examine the very axioms and assumptions from which our models are built. General relativity could not have arisen had not Euclid's parallel axiom (page 136) been first questioned and then abandoned; the mathematics of ordered sets and lattice theory arose as number theory, group theory and logic were stripped down to their common core principles.

For example: the commutative law (Definition 10) almost never applies to language. Word order matters, however successfully 'bag of words' methods have been used in different applications. In building our graph in Chapter 2, we made the simplifying assumption that the relationships we extracted were symmetric, but in the process many phrases such as "chalk and cheese" were used to infer similarity of meaning when they are better viewed as idioms. Methods of combination which allow for non-commutativity (such as the projection mappings of quantum logic) will gain ground over operations which are *always* commutative (such as addition in vector spaces and intersection in set theory), because they don't force language into such an unlikely straitjacket.

As demonstrated in Chapter 7, mathematics still has many simple techniques to offer within the models we already use. For example, the projection mappings used for vector negation can also be used for conjunctions, and their non-commutativity and non-locality may help with situations like the *tiger moth*, where we need to project the compatible features of *tiger* onto the subspace of *moths*, rather than taking the intersection of these subspaces. Another form of composition of vectors, the *tensor product* (Lang, 2002, p. 629), is non-commutative and combines two lower-dimensional (hence less informative) vectors into a single higher-dimensional (and more informative) product. Tensor products have already been used as an ingredient of cognitive and connectionist representations by researchers including Plate (2003), and the use of the tensor product for combining word-meanings has been suggested independently by the quantum physicists Aerts and Czachor (2004). Far from being new, non-commutative product operations on vectors were introduced by Hamilton and Grassmann themselves, when vectors were being invented. The innovation we are suggesting here is that these readily available operations should be applied to new situations, in the noble tradition of mathematical recycling.

Such opportunities exist not only for applying novel operations within the models we use already, but also to whole classes of geometric models which have been developed and used in other areas of science, and are still waiting to be applied to linguistics. For example, Einstein's general theory of relativity rests upon the idea of a *differentiable manifold*, a generalization of the idea of a vector space, freed from some of the Euclidean axioms (Einstein, 1916, Ch. 24). In particular, differentiable manifolds do not presuppose the existence of a global set of coordinates or attributes, and are not constrained to being flat or linear, which (given the relationship between ambiguity and curvature explored in Section 4.5.2) might be a great bonus. Nonlinear techniques involving differentiable manifolds have been used to model some datasets already (Roweis and Saul, 2000), and we should not cling to the flat properties of Euclidean space specifically for the sake of modelling natural language. It doesn't make sense to insist that there should be a unique straight line passing through *fish* which is parallel to the line joining *Scotland* to *piano*!

Manifolds are just the beginning of the standard toolkit of modern differential geometry, which also contains the beautiful mathematics of geodesics (context-sensitive shortest paths in a curved space) and fibre bundles (Ward and Wells, 1990, Ch. 2). In a fibre bundle, each point in the underlying curved manifold or *base-space* can be expanded to give a whole extra space of possibilities localized at that point. For example, the 'curled-up dimensions' of string theory (Greene, 1999, Ch. 8) lie in the fibres of a fibre bundle over the base manifold of spacetime. In a geometric model for language, this would enable a word like *apple* to be represented as a point in the base-space, and a concept such as *fresh green crunchy Granny Smith apple* to be represented using the extra dimensions in the fibre lying over the *apple* point. Such a structure enables more specific information to be represented using extra dimensions, integrating extra detail into a structure of more general knowledge. (Darwin (1859, Ch. 2) noted that a similar process often occurs when a naturalist focusses attention on one particular region or species.)

The dimension of a space is thus intricately bound with the process of measuring the generality or specificity of information. This comes as no surprise: from the point of view of information theory, the dimension (number of coordinates) used to represent a point corresponds exactly to the amount of information needed to specify that point. This correspondence is also used in lattice theory, where the *height* or *dimension* of a node in a taxonomy is determined by the length of the chain of relationships from that node down to the bottom of the taxonomy (Birkhoff, 1967, p. 5).[10] In other words, a postal address taking the form

Number, Street, City, State

contains *four* dimensions of information, because the taxonomy of addresses has four different levels at which the tree can branch. In practice, we rarely describe the world we live on as a 2-dimensional surface — you have almost certainly never given a friend directions to your house by telling them its exact latitude and longitude coordinates and nothing more.

Can mathematics really combine these ideas of points, coordinates, dimensions, decision trees and taxonomies of more general and more specific information, into one harmonious whole? One of the main arguments of this book has been that this is indeed possible, and with

---

[10] A challenge to the mathematically minded readers is to convince themselves that this gives the usual definition of dimensions of vector spaces and their subspaces. A (comparatively easy) chellenge to the historically minded is to find where this definition and correspondence is presented in the writings of (no surprises!) Hermann Grassmann (1862).

the combination of logic and geometry in lattice theory, it has already come to pass.

Can such a mathematics really *work* to combine representations of information from different documents, different knowledge bases, different dictionaries into a coherent whole? Can mathematics really be used to model something as context-dependent as language? In a formalist sense, the answer is probably "no" — not completely. I do not think there will ever be a set of axioms written down that can then be used to look at natural language documents and deduce their meaning using logical steps similar to those used to deduce mathematical theorems. Traditionally, the great proofs of mathematics, such as Euclid's wonderful demonstration that the set of prime numbers must be infinite (*Elements of Geometry*, Bk IX), are *not* context-dependent, they are simply regarded as being true. As we have seen with different geometric models, the mathematical calculus of linguistic concepts, if it is to exist at all, will need to be flexible, robust, and fault-tolerant. It will not predict

> from A and B we deduce P, in all circumstances

but it might be able to suggest

> from A and some combination of B or C, bearing in mind that the speaker's identity is X which suggests that the concepts A, B and C are from the domain Y, it is very probable that P is a valid conclusion, and if you want to be sure, see if you can confirm D or E.

To many mathematicians, this won't even look like proper mathematics, and perhaps it isn't. But if mathematics can contribute at all, it will do so: mathematics actively seeks to involve itself in scientific and technological problems, and when old mathematics has not been appropriate, new mathematics has been discovered. Remember the story of Johannes Kepler (Section 1.3.1), who gave up hope that planetary orbits were described by perfect circles and the Platonic solids, accepting instead that the orbits were elliptical and explaining how the speed of rotation varied. Compared to the classical view that motion in the heavens must be perfect, and therefore perfectly circular (for example, see Aristotle, *On the Heavens*, Bk 1), the idea that planetary orbits could come in an infinite variety of squashed elliptical shapes and sizes was contrary to the beauty of mathematics, and yet paved the way for the theory of gravitation (see Bohm, 1980, Ch. 5). If there is to be a mathematics of human communication, its greatness will not lie in the elegance with which it imposes purity on messy subject matter, but in the plastic adaptability of simple abstractions, serving as an unifying description for superficially disparate phenomena.

We turn for one last time to the story of quantum theory in the twentieth century. Albert Einstein, one of the main contributors, once wrote "The more success the quantum theory has, the sillier it looks."[11] But in the hands of mathematicians like Paul Dirac (1930) and John von Neumann (1932), a new theory of geometric operators and logical combinations in vector spaces was born. Instead of predicting the exact outcome of an observation, the theory predicted exactly the *probability* of different outcomes, opening a new era in the attitude of physics to the doctrine of determinism, and motivating some wonderful new mathematics which we have only recently applied to concepts in language. If physics can accept that its laws do not predict the universe exactly, perhaps mathematics can find the broadness of heart to accept, as its own, theories which depend on experience and context.

Epilogue — Aristotle the mathematician

To those who have finally reached this epilogue, many congratulations and thanks! As well as finding out lots of exciting information about language technology in the present day, you have encountered many stories about the scientists whose discoveries over the centuries have led us to this point. Often, this history goes untold in technical works because it's not considered relevant: nonetheless, I hope the emphasis on history and the way it brings the present into focus has made this journey more valuable, not to mention more enjoyable. All of our heroes — Euclid, Descartes, George Boole, Charles Darwin, Newton and Einstein, Hamilton and Grassmann, Birkhoff and von Neumann, and many more — they learned from the past, turning not only to the recent books and papers of their own generations, but old stones left unturned by many others for years, concealing treasures only from those who did not seek them. If this book encourages any readers to encounter great scientific discoveries first-hand by turning to the words of their discoverers, that alone would make the effort worthwhile to me.

---

[11] Letter to Heinrich Zangger, May 20, 1912.

One of the dangers of reading only up-to-the-minute works and ignoring primary sources is that it reinforces prejudice about people we think we know *about*, but do not *know*. My school textbooks in mechanics and biology mentioned Aristotle only to ridicule his ideas of force, movement, and the generation of species, thus demonstrating the relative superiority of scientists like Galileo, Newton, Pasteur and Darwin. No mention ever was made of the way Aristotle's work laid the foundations for these followers, and often had to be changed only a little to accommodate later developments. Another piece of received wisdom which I often encountered, and for many years failed to question, is that Aristotle wasn't much of a mathematician, and wasn't very well-disposed towards the discipline.

This preliminary contact left me with no motivation to read Aristotle's works for myself, and as I set out to write this book it was with the intention of offering a mathematical *alternative* to the Aristotelean tradition (just as Darwin was originally hired for the Beagle voyage to refute any of the dangerous evolutionary ideas that were beginning to surface at the time — fortunately for science, he kept his eyes and his mind open). I never thought to find myself writing a book on geometry in which Aristotle turned out to be the greatest hero, but to my surprise and eventual joy, as I began checking a few of his writings, starting with the *Categories*, I found that my work was already full of his enormous influence and unsung mathematical discoveries, leaving me to conclude that Aristotle was not only a great physical scientist (not to mention philosopher and linguist), but also a great mathematician. Aristotle's greatness in the physical sciences is universally accepted by all who read his work, and in spite of my school biology textbook, will ever remain so. But the claim that he was also a great *mathematician* is so contrary to received wisdom that it requires substantiation.

The main negative evidence cited against Aristotle as a mathematician is the lack of accurate measurement in his work: but as we saw in Chapter 1, the making of measurements is not a prerequisite for developing and using mathematics, since we are often primarily interested in *relationships*, and often numbers are useful only insofar as they enable us to compare the relative importance of relationships. If a lack of measurement does not disqualify one from being a great scientist, it certainly does not disqualify one from being a great mathematician.

So much for the negative: now the positive evidence. First of all, the use of symbols to represent concepts was introduced by Aristotle (very noticeably, between the first and second chapters of the *Prior Analytics*). The technique had been applied to geometry by Euclid within a couple of generations, though it was notably absent from algebra itself for centuries: nonetheless, the substitution of a letter to stand for any mathematical object is not due to Descartes, it is due to Aristotle. The enormous usefulness of this contribution alone would rank anyone as a great mathematician.

Secondly, Aristotle's *use* of mathematics, particularly geometry, is impeccable, given the mathematics of his day (see for example the *Physics* and *On the Heavens*). While some of his leaps from theoretical geometry to physical conclusions are fanciful, his mathematical reasoning is correct and professional. In this respect, Aristotle may be likened to Albert Einstein: Einstein did not make great contributions to pure mathematics, but was a great *historian* of mathematics, finding exactly those parts he needed to frame his scientific developments, even if they were buried in obscure and unfashionable writings from the previous century.

Thirdly, much of the mathematical theory behind ordered sets, taxonomies and lattices was introduced by Aristotle, centuries ahead of its time. Consider, as a single example, the discussion of chains of predicates in the *Posterior Analytics* (Bk. 1, Ch. 19). In just one and a half pages packed with pure mathematics, Aristotle introduces the idea of immediate predication (*covering* in the terminology of contemporary ordered set theory (Davey and Priestley, 1990, p. 11)), the possibility of reasoning through middle terms which are credible but not real (anticipating some of the questions of intuitionistic logic (Gamut, 1991, p. 140)), the notion of top and bottom terms in a concept hierarchy (introduced again by Boole (1847)), and analyzes the viability of infinite chains in such a hierarchy. The sophistication of this chapter alone would have been remarkable in any mathematical work dated before the middle of the 19th century: to find it in a work predating this by over two thousand years is astonishing. It seems that Aristotle has not been considered a great mathematician because by the time mathematics was ready finally to inherit his ideas, his name had long since slipped from the reckoning, leaving many of the laurels to his distant successors.

CONCLUSIONS AND CURTAIN CALL / 303

Finally, the union of geometry and logic, pursued in this book as a means of adding implication to geometrical representations and empirical breadth and flexibility to logical ones, begins with recognizing the equivalence between logical implication and geometric containment — introduced by Aristotle (*Prior Analytics*, Bk 1, Ch. 1). Our intuitive awareness of the significance of geometric representations goes back to prehistory, but in this great step, it was Aristotle who began to build a systematic bridge between geometry and meaning.

References

Abbott, E. 1884. *Flatland*. Dover edition 1992.

Aerts, D. and M. Czachor. 2004. Quantum aspects of semantic analysis and symbolic artificial intelligence. *J. Phys. A: Math. Gen.* 37:L123–L132.

Agirre, E. and G. Rigau. 1996. Word sense disambiguation using conceptual density. In *Proceedings of the 16th International Conference on Computational Linguistics (COLING-96)*, pages 16–22. Copenhagen, Denmark.

Aitchison, J. 2002. *Words in the Mind: An Introduction to the Mental Lexicon*. Blackwell, 3rd edn.

Baeza-Yates, R. and B. Ribiero-Neto. 1999. *Modern Information Retrieval*. Addison Wesley / ACM Press.

Baldwin, T., C. Bannard, T. Tanaka, and D. Widdows. 2003. An empirical model of multiword expression decomposability. In *Proceedings of the ACL-03 Workshop on Multiword Expressions: Analysis, Acquisition and Treatment*. Sapporo, Japan.

Barnes, J. 2001. *Aristotle: A Very Short Introduction*. Oxford University Press.

Bean, C. A. and R. Green, eds. 2001. *Relationships in the Organization of Knowledge*. Kluwer.

Berry, M. W., S. T. Dumais, and G. W. O'Brien. 1995. Using linear algebra for intelligent information retrieval. *SIAM Review* 37(4):573–595.

Bertino, E., B. C. OOI, R. Sacks-Davis, K.-L. Tan, J. Zobel, B. Shidlovsky, and B. Catania. 1997. *Indexing Techniques for Advanced Database Systems*. Kluwer.

Birkhoff, G. 1967. *Lattice theory*. American Mathematical Society, 3rd edn. First edition 1940.

Birkhoff, G. and J. von Neumann. 1936. The logic of quantum mechanics. *Annals of Mathematics* 37:823–843.

Bodenreider, O. and R. Green. 2001. Relationships among knowledge structures: Vocabulary integration within a subject domain. In Bean and Green (2001), chap. 6, pages 81–98.

Bohm, D. 1951. *Quantum Theory*. Prentice-Hall. Republished by Dover, 1989.

Bohm, D. 1980. *Wholeness and the Implicate Order*. Routledge Classics, republished 2002. Routledge.

Bollobás, B. 1998. *Modern Graph Theory*. No. 184 in Graduate Texts in Mathematics. Springer-Verlag.

Boole, G. 1847. *The Mathematical Analysis of Logic*. Macmillan. Republished by St Augustine's press, 1998, introduction by John Slater.

Boole, G. 1854. *An Investigation of the Laws of Thought*. Macmillan. Dover edition, 1958.

Boyer, C. B. and U. C. Merzbach. 1991. *A History of Mathematics*. Wiley, 2nd edn.

Brants, T., F. Chen, and I. Tsochantaridis. 2002. Topic-based document segmentation with probabilistic latent semantic analysis. In *Conference on Information and Knowledge Management (CIKM)*, pages 211–218.

Brin, S. and L. Page. 1998. The anatomy of a large-scale hypertextual Web search engine. *Computer Networks and ISDN Systems* 30(1–7):107–117.

Budanitsky, A. and G. Hirst. 2001. Semantic distance in WordNet: An experimental, application-oriented evaluation of five measures. In *NAACL Workshop on WordNet and Other Lexical Resources*. Pittsburgh, Pennsylvania.

Buitelaar, P. 1998. *CoreLex: Systematic Polysemy and Underspecification*. Ph.D. thesis, Computer Science Department, Brandeis University.

Bush, R. R. and F. Mosteller. 1951. A model for stimulus generalization and discrimination. *Psychological Review* 48:413–423.

Callan, J. P., W. B. Croft, and S. M. Harding. 1992. The INQUERY retrieval system. In *Proceedings of DEXA-92, 3rd International Conference on Database and Expert Systems Applications*, pages 78–83.

Cantor, G. 1895. *Contributions to the Founding of the Theory of Transfinite Numbers*. Repinted by Dover, 1952, introduction by Philip Jourdain.

Caraballo, S. 1999. Automatic construction of a hypernym-labeled noun hierarchy from text. In *37th Annual Meeting of the Association for Computational Linguistics (ACL99)*, pages 120–126.

Card, S. K., J. D. Mackinlay, and B. Schneiderman, eds. 1999. *Readings in Information Visualization*. Morgan Kaufmann.

Carpenter, B. 1992. *The Logic of Typed Feature Structures*. Cambridge University Press.

Carroll, L. 1865. *Alice's Adventures in Wonderland*. Penguin Popular Classics (1994).

Carroll, L. 1872. *Through the Looking Glass*. Penguin Popular Classics (1994).

Cederberg, S. and D. Widdows. 2003. Using LSA and noun coordination information to improve the precision and recall of automatic hyponymy extraction. In *7th Conference on Natural Language Learning (CoNNL-03)*. Edmonton, Canada.

Chakrabarti, S., B. Dom, R. Agrawal, and P. Raghavan. 1998. Scalable feature selection, classification and signature generation for organizing large text databases into hierarchical topic taxonomies. *The VLDB Journal (Very Large Data Bases)* 7(3):163–178.

Chang, B. S. W., K. Jönsson, M. A. Kazmi, M. J. Donoghue, and T. P. Sakmar. 2002. Recreating a functional ancestral archosaur visual pigment. *Mol. Biol. Evol.* 19:1483–1489.

Chappelier, J.-C., M. Rajman, R. Aragües, and A. Rozenknop. 1999. Lattice parsing for speech recognition. In *Le Tratement Automatique des Langues Naturelles (TALN99)*. Cargèse, Corsica.

Chartrand, G. 1985. *Introductory Graph Theory.* Dover.

Cohen, D. 1989. *An introduction to Hilbert Spaces and Quantum Logic.* Spriger-Verlag.

Collins, A. M. and M. R. Quillian. 1969. Retrieval time from semantic memeory. *Journal of verbal learning and verbal behaviour* 8:240–247.

Coxeter, H. S. M. 2003. *Projective Geometry.* Springer-Verlag.

Cruse, A. 2000. *Meaning in Language: An Introduction to Semantics and Pragmatics.* Oxford Textbooks in Linguistics. Oxford University Press.

Cruse, D. A. 2002. Hyponymy and its varieties. In R. Green, C. Bean, and S. H. Myaeng, eds., *The Semantics of Relationships: An interdisciplinary perspective*, chap. 1. Kluwer.

Daniélou, A. 1943. *Music and the Power of Sound: the influence of tuning and pitch on consciousness.* Inner Traditions International. Revised edition 1995.

Darwin, C. 1859. *The Origin of Species.* Penguin Classics 1982, ed. J. W. Burrows.

Davey, B. A. and H. A. Priestley. 1990. *Lattices and Order.* Cambridge University Press.

Deerwester, S., S. Dumais, G. Furnas, T. Landauer, and R. Harshman. 1990. Indexing by latent semantic analysis. *Journal of the American Society for Information Science* 41(6):391–407.

Delancey, S. 2000. The universal basis of case. *Logos and Language* 1(2).

Descartes, R. 1637. *The Geometry of René Descartes.* Dover edition (1954) with facsimile of the original.

Dirac, P. 1930. *The principles of quantum mechanics.* Clarendon Press, Oxford, 4th edn.

Dolan, W., L. Vanderwende, and S. Richardson. 1993. Automatically deriving structured knowledge bases from online dictionaries. In *Pacific Assoc. for Computational Linguistics (PACLING 93)*, pages 5–14.

Dorow, B. and D. Widdows. 2003. Discovering corpus-specific word-senses. In *Conference Companion, 10th Conference of the European Chapter of the Association for Computational Linguistics (EACL-03)*, pages 79–82. Budapest, Hungary.

Dumais, S. and H. Chen. 2000. Hierarchical classification of web content. In *Proceedings of the 23rd annual conference on Research and Development in Information Retrieval (SIGIR-00)*, pages 256–263. Athens, Greece.

Dunlop, M. 1997. The effect of accessing nonmatching documents on relevance feedback. *ACM Transactions on Information Systems* 15(2):137–153.

Eckmann, J.-P. and E. Moses. 2002. Curvature of co-links uncovers hidden thematic layers in the world-wide web. *Proceedings of the National Academy of Science* 99:5825–5829.

Einstein, A. 1916. *Relativity: the Special and General Theory*. Holt and Company (English edition, 1920). Republished by Dover, 2001.

El-Hoshy, L. 2001. Relationships in library of congress subject headings. In Bean and Green (2001), chap. 6, pages 81–98.

Evens, M. W., ed. 1988. *Relational Model of the Lexicon: Representing Knowledge in Semantic Networks*. Cambridge University Press.

Faatz, A., S. Hoermann, C. Seeberg, and R. Steinmetz. 2001. Conceptual enrichment of ontologies by means of a generic and configurable approach. In *Proceedings of the European Summer School on Logic, Language and Information (ESSLLI-01) Workshop on Semantic Knowledge Acquisition and Categorisation*.

Fellbaum, C. 1998a. A semantic network of English verbs. In Fellbaum (1998b), chap. 3, pages 69–104.

Fellbaum, C., ed. 1998b. *WordNet: An Electronic Lexical Database*. MIT Press.

Foskett, D. J. 1997. Thesaurus. In K. S. Jones and P. Willett, eds., *Readings in Information Retrieval*, pages 111–134. Morgan Kaufmann.

Frege, G. 1884. *The Foundations of Arithmetic (1884)*. Blackwell edition, 1974, translated by J. L. Austin.

Fuller, B. 1975. *Synergetics: Explorations in the Geometry of Thinking*. Macmillan / Collier. in collaboration with E. J. Applethwaite.

Gabora, L. and D. Aerts. 2002. Contextualizing concepts using a mathematical generalization of the quantum formalism. *Journal of Experimental and Theoretical Artificial Intelligence* 14:327–358.

Gamut, L. 1991. *Logic, Language, and Meaning*. University of Chicago Press.

Ganter, B. and R. Wille. 1999. *Formal Concept Analysis: Mathematical Foundations*. Springer.

Gärdenfors, P. 2000. *Conceptual Spaces: The Geometry of Thought*. Bradford Books MIT Press.

Goscinny, R. and A. Uderzo. 1974. *Asterix et les Goths*. Editeur Dargaud.

Grassmann, H. 1862. *Extension Theory*. History of Mathematics Sources. American Mathematical Society, London Mathematical Society, 2000. translated by Lloyd C. Kannenberg.

Greene, B. 1999. *The Elegant Universe*. Norton.

Grefenstette, G. 1994. *Explorations in Automatic Thesaurus Discovery*. Kluwer.

Grefenstette, G., ed. 1998. *Cross-language information retrieval*. Kluwer.

Guillemin, V. 1968. *The Story of Quantum Mechanics*. Charles Scribner's Sons. Republished by Dover, 2003.

Guthrie, L., J. Pustejovsky, Y. Wilks, and B. Slator. 1996. The role of lexicons in natural language processing. *Communications of the Associate of Computational Machinery (ACM)* 39(1):63–72.

Hamilton, A. G. 1982. *Numbers, Sets and Axioms*. Cambridge University Press.

Hamilton, S. W. R. 1847. On quaternions. *Proc. Royal Irish Acad.* 3:1–16.

Hamming, R. W. 1980. *Coding and Information Theory*. Prentice-Hall.

Hankerson, D. R., D. G. Hoffman, D. A. Leonard, C. C. Lindner, K. T. Phelps, C. A. Rodger, and J. R. Wall. 2000. *Coding Theory and Cryptography: the Essentials*. Marcel Dekker.

Harry, B. 1992. *The Ultimate Beatles Encyclopedia*. MJF Books.

Hausdorff, F. 1914. *Grundzüge der Mengenlehre*. von Veit (Germany), 1914. Republished as *Set Theory*, 2nd ed. Chelsea (New York), 1962.

Hearst, M. 1999. User interfaces and visualization. In Baeza-Yates and Ribiero-Neto (1999), chap. 10, pages 257–324.

Hearst, M. and H. Schütze. 1993. Customizing a lexicon to better suit a computational task. In *Proceedings of the Special Interest Group on the Lexicon, Association for Computational Linguistics (ACL-SIGLEX Workshop)*. Columbus, Ohio.

Hearst, M. A. 1992. Automatic acquisition of hyponyms from large text corpora. In *Proceedings of the 14th International Conference on Computational Linguistics (COLING-92)*. Nantes, France.

Hearst, M. A. 1998. WordNet: An electronic lexical database. In Fellbaum (1998b), chap. 5, Automated discovery of WordNet relations, pages 131–152.

Heath, T. L., ed. 1956. *The Thirteen Books of Euclid's Elements*, vol. I-III. Dover.

Hersh, W., C. Buckley, T. J. Leone, and D. Hickam. 1994. Ohsumed: An interactive retrieval evaluation and new large test collection for research. In *Proceedings of the 17th annual conference on Research and Development in Information Retrieval (SIGIR-94)*, pages 192–201.

Herskovits, A. 1986. *Language and Spatial Cognition*. Cambridge University Press.

Hirst, G. and D. St-Onge. 1998. Lexical chains as representations of context for the detection and correction of malapropisms. In Fellbaum (1998b), chap. 13, pages 305–332.

Hirvensalo, M. 2001. *Quantum Computing*. Springer.

Hofmann, T. 1999. Probabilistic latent semantic analysis. In *Uncertainty in Artificial Intelligence (UAI'99)*. Stockholm, Sweden.

Horn, L. 2001. *A Natural History of Negation*. CSLI publications.

Huddleston, R. and G. K. Pullum. 2002. *The Cambridge Grammar of the English Language*. Cambridge University Press.

Jammer, M. 1993. *Concepts of Space: The History of Theories of Space in Physics*. Dover, 3rd edn.

Jänich, K. 1994. *Linear algebra*. Undergraduate Texts in Mathematics. Springer-Verlag.

Janssen, M. 2002. *SIMuLLDA: a Multilingual Lexical Database Application using a Structured Interlingua*. Ph.D. thesis, Utrecht University.

Jiang, J. and D. Conrath. 1997. Semantic similarity based on corpus statistics and lexical taxonomy. In *Proceedings of International Conference on Research in Computational Linguistics*. Taiwan.

Jurafsky, D. and J. H. Martin. 2000. *Speech and Language Processing*. New Jersey: Prentice Hall.

Kacmarcik, G., C. Brockett, and H. Suzuki. 2000. Robust segmentation of Japanese text into a lattice for parsing. In *Proceedings of the 18th International Conference on Computational Linguistics (COLING-00)*, pages 390–6. Saarbrücken, Germany.

Kilgarriff, A. 1993. Dictionary word sense distinctions: An enquiry into their nature. *Computers and the Humanities* 26(1–2):365–387.

Kleinberg, J., S. Kumar, P. Raghavan, S. Rajagopalan, and A. Tomkins. 1999. The web as a graph: Measurements, models and methods. In *Invited survey at the International Conference on Combinatorics and Computing*.

Kleinberg, J. and S. Lawrence. 2001. The structure of the web. *Science* 294:1849–1850.

Kowalski, G. 1997. *Information retrieval systems: theory and implementation*. Kluwer academic publishers.

Landauer, T. and S. Dumais. 1997. A solution to Plato's problem: The latent semantic analysis theory of acquisition. *Psychological Review* 104(2):211–240.

Lang, S. 2002. *Algebra*. No. 211 in Graduate Texts in Mathematics. Springer-Verlag.

Lasenby, J., A. Lasenby, and C. Doran. 2000. A unified mathematical language for physics and engineering in the 21st century. *Phil. Trans. R. Soc. Lond. A* 358:21–39.

Leacock, C. and M. Chodorow. 1998. WordNet: An electronic lexical database. In Fellbaum (1998b), chap. 11, pages 265–283.

Leech, G., R. Garside, and M. Bryant. 1994. Claws4: The tagging of the British National Corpus. In *Proceedings of the 15th International Conference on Computational Linguistics (COLING-94)*, pages 622–628. Kyoto, Japan.

Lehmann, F., ed. 1992. *Semantic Networks in Artificial Intelligence*. Pergamon Press.

Lehmann, W. P. 1993. *Historical Linguistics*. Routledge, 3rd edn.

Lenat, D. B. and R. V. Guha. 1990. *Building Large Knowledge-Based Systems: Representation and Inference in the Cyc Project*. Addison-Wesley.

Li, H. and N. Abe. 1998a. Generalizing case frames using a thesaurus and the MDL principle. *Computational Linguistics* 24(2):217–244.

Li, H. and N. Abe. 1998b. Word clustering and disambiguation based on co-occurence data. In *Proceedings of the 17th International Conference on Computational Linguistics / 38th Annual Meeting of the Association for Computational Linguistics (COLING-ACL-98)*, pages 749–755. Montreal, Canada.

Li, X., R. Singh, and R. M. Stern. 2002. Lattice combination for improved speech recognition. In *Proc. of the International Conference of Spoken Language Processing*. Denver, USA.

Lin, D. 1998a. Automatic retrieval and clustering of similar words. In *Proceedings of the 17th International Conference on Computational Linguistics / 38th Annual Meeting of the Association for Computational Linguistics (COLING-ACL-98)*. Montreal, Canada.

Lin, D. 1998b. An information-theoretic definition of similarity. In *Proceedings of the 15th International Conference on Machine Learning (ICML-98)*. Madison, WI.

Lin, D. and P. Pantel. 2002. Concept discovery from text. In *Proceedings of 19th International Conference on Computational Linguistics (COLING-02)*, pages 577–583. Taipei, Taiwan.

Littman, M. L., S. T. Dumais, and T. K. Landauer. 1998. Automatic cross-linguage information retrieval using latent semantic indexing. In G. Grefenstette, ed., *Cross-language information retrieval*, chap. 4. Kluwer.

Mann, G. and D. Yarowsky. 2003. Unsupervised personal name disambiguation. In *7th Conference on Natural Language Learning (CoNNL-03)*. Edmonton, Canada.

Manning, C. D. and H. Schütze. 1999. *Foundations of Statistical Natural Language Processing*. The MIT Press.

Marcus, M. P., B. Santorini, and M. A. Marcinkiewicz. 1993. Building a large annotated corpus of English: the Penn treebank. *Computational Linguistics* 19(2):313–30.

McKeon, R., ed. 1941. *The Basic Works of Aristotle*. Random House.

Melamed, I. D. 2000. Pattern recognition for mapping bitext correspondence. In J. Véronis, ed., *Parallel Text Processing*, pages 25–48. Kluwer.

Mervis, C. and E. Rosch. 1981. Categorization of natural objects. *Annual Review of Psychology* 32:89–115.

Miller, G. A. 1998a. Nouns in WordNet. In Fellbaum (1998b), chap. 1, pages 23–46.

Miller, G. A. and W. G. Charles. 1991. Contextual correlates of semantic similarity. *Language and Cognitive Processes* 6(1):1–28.

Miller, K. J. 1998b. Modifiers in WordNet. In Fellbaum (1998b), chap. 2, pages 47–68.

Mitchell, T. 1997. *Machine Learning*. McGraw-Hill.

Miyamoto, S. 1990. *Fuzzy sets in information retrieval and cluster analysis*. Kluwer.

Partee, B. H., A. ter Meulen, and R. E. Wall. 1993. *Mathematical Methods in Linguistics*. Kluwer.

Patwardhan, S., S. Banerjee, and T. Pedersen. 2002. Using semantic relatedness for word sense disambiguation. In *Proceedings of the Fourth International Conference on Intelligent Text Processing and Computational Linguistics*. Mexico City.

Pereira, F., N. Tishby, and L. Lee. 1993. Distributional clustering of english words. In *30th Annual Meeting of the Association for Computational Linguistics (ACL-93)*, pages 183–190. Columbus, Ohio.

Plate, T. 2003. *Holographic Reduced Representations: Distributed Representation for Cognitive Structures*. CSLI Publications.

Polkinghorne, J. 2002. *Quantum Theory, A Very Short Introduction*. Oxford University Press.

Pollard, C. and I. Sag. 1994. *Head-Driven Phrase Structure Grammar*. University of Chicago Press.

Proctor, P., ed. 1978. *The Longman Dictionary of Contemporary English (LDOCE).* London: Longman.
Pullman, B. 1998. *The atom in the history of human thought.* Oxford University Press.
Pustejovsky, J. 1995. *The Generative Lexicon.* MIT press.
Putnam, H. 1976. The logic of quantum mechanics. In *Mathematics, Matter and Method*, pages 174–197. Cambridge University Press.
Quillian, M. R. 1968. Semantic memory. In M. Minsky, ed., *Semantic Information Processing*, chap. 4, pages 227–270. MIT press.
Resnik, P. 1999. Semantic similarity in a taxonomy: An information-based measure and its application to problems of ambiguity in natural language. *Journal of artificial intelligence research* 11:93–130.
Riloff, E. and J. Shepherd. 1997. A corpus-based approach for building semantic lexicons. In *Proceedings of the Second Conference on Empirical Methods in Natural Language Processing (EMNLP-97)*, pages 117–124. Association for Computational Linguistics.
Roark, B. and E. Charniak. 1998. Noun-phrase co-occurence statistics for semi-automatic semantic lexicon construction. In *Proceedings of the 17th International Conference on Computational Linguistics / 38th Annual Meeting of the Association for Computational Linguistics (COLING-ACL-98)*, pages 1110–1116.
Rooth, M., S. Riezler, D. Prescher, G. Carroll, and F. Beil. 1999. Inducing a semantically annotated lexicon via em-based clustering. In *In Proceeding of the 37th Annual Meeting of the Association for Computational Linguistics (ACL-99).* College Park, Maryland.
Rosch, E. 1975. Cognitive representations of semantic categories. *Journal of Experimental Psychology: General* 104:192–233.
Roweis, S. and L. Saul. 2000. Nonlinear dimensionality reduction by locally linear embedding. *Science* 290(5500):2323–2326.
Salton, G. and C. Buckley. 1990. Improving retrieval performance by relevance feedback. *Journal of the American society for information science* 41(4):288–297.
Salton, G. and M. McGill. 1983. *Introduction to modern information retrieval.* McGraw-Hill.
Schütze, H. 1997. *Ambiguity resolution in language learning.* CSLI Publications.
Schütze, H. 1998. Automatic word sense discrimination. *Computational Linguistics* 24(1):97–124.
Shannon, C. 1948. A mathematical theory of communication. *Bell system technical journal* 27:379–423, 623–656.
Sigman, M. and G. A. Cecchi. 2002. The global organization of the wordnet lexicon. *Proceedings of the National Academy of Sciences of the USA (PNAS)* 99(3):1742–1747.
Smith, D. E., ed. 1929. *A Source Book in Mathematics.* McGraw Hill. Dover edition, 1959.
Sowa, J. F. 2000. *Knowledge Representation: Logical, Philosophical, and Computational Foundations.* Brooks Cole Publishing.

Sparck Jones, K. 1986. *Synonymy and Semantic Classification*. Edinburgh University Press. (Originally Cambridge PhD thesis, 1964).

Sparck Jones, K. and P. Willett, eds. 1997. *Readings in Information Retrieval*. Morgan Kaufmann.

Squires, E. 1986. *The Mystery of the Quantum World*. Adam Hilger.

Stumme, G. and A. Maedche. 2001. FCA-MERGE: Bottom-up merging of ontologies. In *17th International Joint Conference on Artificial Intelligence (IJCAI-01)*, pages 225–234.

Tenenbaum, J. B., V. de Silva, and J. C. Langford. 2000. A global geometric framework for nonlinear dimensionality reduction. *Science* 290(5500):2319–2323.

Trefethen, L. N. and D. Bau. 1997. *Numerical Linear Algebra*. Society for Industrial and Applied Mathematics (SIAM).

Trudeau, R. J. 1994. *Introduction to Graph Theory*. Dover.

Turtle, H. and W. B. Croft. 1989. Inference networks for document retrieval. In *Proceedings of the 13th annual conference on Research and Development in Information Retrieval (SIGIR-89)*, pages 1–24.

Tversky, A. 1977. Features of similarity. *Psychological Review* 84(4):327–352.

Tversky, A. 1982. Similarity, separability and the triangle inequality. *Psychological Review* 89(2):123–154.

Vallejo, R. J. 1993. *Linear algebra: an introduction to abstract mathematics*. Undergraduate Texts in Mathematics. Springer-Verlag.

Van der Waerden, B. L. 1985. *A History of Algebra*. Springer-Verlag.

van Dongen, S. 2000. *Graph Clustering by Flow Simulation*. Ph.D. thesis, University of Utrecht.

Varadarajan, V. S. 1985. *Geometry of Quantum Theory*. Springer-Verlag.

von Neumann, J. 1932. *Mathematical Foundations of Quantum Mechanics*. Princeton Univ Press. Reprinted 1996.

Vossen, P. 1998. Introduction to EuroWordNet. *Computers and the Humanities* 32(2-3):73–89.

Ward, R. S. and R. O. Wells. 1990. *Twistor Geometry and Field Theory*. Cambridge University Press.

Wheeler, J. A. 1991. Albert Einstein. In T. Ferris, ed., *The World Treasury of Physics, Astronomy and Mathematics*, pages 563–576. Little, Brown and Co.

Widdows, D. 2003a. A mathematical model for context and word-meaning. In *Fourth International and Interdisciplinary Conference on Modeling and Using Context (CONTEXT-03)*. Stanford, California.

Widdows, D. 2003b. Orthogonal negation in vector spaces for modelling word-meanings and document retrieval. In *Proceedings of the 41st Annual Meeting of the Association for Computational Linguistics (ACL-03)*. Sapporo, Japan.

Widdows, D. 2003c. Unsupervised methods for developing taxonomies by combining syntactic and statistical information. In *Proceedings of Human Langauge Technology / North American Chapter of the Association for Computational Linguistics (HLT-NAACL-03)*. Edmonton, Canada.

Widdows, D., S. Cederberg, and B. Dorow. 2002a. Visualisation techniques for analysing meaning. In *Fifth International Conference on Text, Speech and Dialogue (TSD-02)*, Lecture Notes in Artificial Intelligence 2448, pages 107–115. Brno, Czech Republic: Springer.

Widdows, D. and B. Dorow. 2002. A graph model for unsupervised lexical acquisition. In *19th International Conference on Computational Linguistics (COLING-02)*, pages 1093–1099. Taipei, Taiwan.

Widdows, D., B. Dorow, and C.-K. Chan. 2002b. Using parallel corpora to enrich multilingual lexical resources. In *Third International Conference on Language Resources and Evaluation (LREC-02)*, pages 240–245. Las Palmas, Spain.

Widdows, D. and M. Higgins. 2004. Geometric ordering of concepts, logical disjunction, learning by induction, and spatial indexing. In *Compositional Connectionism in Cognitive Science*. AAAI Fall Symposium Series, Washington, DC.

Widdows, D. and S. Peters. 2003. Word vectors and quantum logic. In *Proceedings of the Eighth Mathematics of Language Conference (MoL8)*. Bloomington, Indiana.

Wierzbicka, A. 1996. *Semantics: Primes and Universals*. Oxford University Press.

Wilce, A. 2003. Quantum logic and probability theory. In E. N. Zalta, ed., *The Stanford Encyclopedia of Philosophy (Spring 2003 Edition)*. Stanford University.

Witten, I. H., A. Moffat, and T. C. Bell. 1999. *Managing Gigabytes: Compressing and Indexing Documents and Images*. Morgan Kaufmann, 2nd edn.

Wittgenstein, L. 1953. *Philosophical Investigations*. Blackwell: Blackwell. 3rd edition, 2001.

Yang, Y., J. Carbonell, R. Brown, and R. Frederking. 1998. Translingual information retrieval: Learning from bilingual corpora. *Artificial Intelligence Journal special issue: Best of IJCAI-97* pages 323–345.

Zalta, E. N. 2003. Frege's logic, theorem, and foundations for arithmetic. In E. N. Zalta, ed., *The Stanford Encyclopedia of Philosophy (Spring 2003 Edition)*. Stanford University.

Index

Relationship Symbols

$<$, ordering of numbers by magnitude, 14, 116, 248
$\leq$, ordering of numbers by magnitude, 14, 76, 248
$\leftrightarrow$, link in graph, 120
$\preceq$, ordering in ordered set, 249
$\subseteq$, set containment, 13, 263
  see also implication, 16
$\leftrightarrow$, link in graph, 51
$\rightarrow$, link in directed graph, 73
$\sqsubseteq$, link in hierarchy, 76, 273
  see also hyponymy, 82

Number Symbols

$\mathbb{C}$, complex numbers, 34
$\mathbb{N}$, natural numbers, 18–24, 33
$\mathbb{Q}$, rational numbers, 34
  see also ratios, 24
$\mathbb{R}$, real numbers, 28, 138
$\mathbb{R}^2$, the plane, 140
$\mathbb{R}^n$, real vector space of dimension n, 151, 158
$\mathbb{Z}$, integers (Zahlen), 34
$\mathbb{Z}_2$, binary numbers, 205

Operator Symbols

$+$, addition of vectors, 134–138
$+$, vector sum of subspaces, 236
Σ, summation of several numbers, 153
$\cap$, set intersection, 14
  and lattice meet, 263
$\cdot$, scalar product, 152, 210
$\cup$, set union, 14
  and lattice join, 263
cos, cosine similarity measure, 106, 158–160
$\perp$, perpendicular subspace, 281
$\times$, Cartesian product, 50
$\vee$, lattice join, 260
  and set union, 263
$\wedge$, lattice meet, 260
  and set intersection, 263
$'$, set complement, 14
AND, $\wedge$, conjunction, 206
  and lattice meet, 278
  and set intersection, 207
NOT, $\neg$, negation, 206
  see also complement, 279
OR, $\vee$, disjunction, 206
  and lattice join, 278
  and set union, 207

INDEX / 315

acyclic graph, 254
affinity score function, 89
ambiguity, 62–65, 83, 84
   and measuring distance, 109
   and transitive relationships, 118
   systematic, 128, 196, 219
   translational, 193
antisymmetric relationships, *see*
   relationships, antisymmetric
antonym, 84, *see also* contrary, 286
Appolonius, 140
Aristotle, 17, 33, 70, 300–303
   and the quantum, 216
   logic, *see* logic, Aristotelean
arrays, 151
axioms
   metric space, 100
   of geometry, 8, 136, 138
   of logic, 281
   ordered set, 250
   vector space, 139

Babylonian number system, 24
basis, 142
bilingual applications, 189–193
binary values, 205
Boole, George, 202–207
Boolean logic, 205–208
   and lattice theory, 283
Boolean search engine, 207
bounding box, 225, *see also*
   Manhattan metric, 241
British National Corpus (BNC), 45, 168

Cantor, Georg, 39
Cartesian product, 50
case roles, 4
Chinese numerology, 20
circle, 226
   unit, 105
class labelling, 88–94
   and WORDSPACE, 194–196
   and join or disjunction, 254
clustering, 118, 179–183
   incremental, 122

commutative, 296
compactness, 234
complement, 14, *see also* negation, 279–283
   Boolean, 284
   in quantum logic, 281
   orthogonal, 281
complex numbers, 34
composition, 161, 191
conductance, 104
conjunction
   logical, 206
connectionism, 12
containment, 13, 16, 224, 250, 274
content-bearing words, 173
continuous quantities, 27, 31
   and information retrieval, 163
   and logic, 200
contrary, 84, 256, 269, *see also*
   complement, 286
cooccurrence, 173
coordinates, 138
   Cartesian, 140
coordination patterns, 55
corpora, 45–47
cosine function, 106
cosine similarity measure, 105–107
   and Euclidean distance, 159
   in $\mathbb{R}^n$, 158
   in WORDSPACE, 171
counting, 26, 35
cross-table, 269, 271
curvature, 123, 135
cycle, 63
cycle in graph, 253

Darwin, Charles, 70
data sparseness, 118
De Morgan Law, 238
De Morgan law, 206
demonstration
   bilingual, 189
   word spectra, 183
   WORDSPACE, 170
demonstrations
   negation in WORDSPACE, 214

Descartes, René, 140, 164
differentiae, 80
dimension, 225
dimensions, 3, 92, 142–144
   reducing, 175
Dirac, Paul, 220
directed acyclic graph (DAG), 251
directed graph, 73
directed link, 73
direction, 105
directories or folders, 78
discrete quantities, 31, 163
   and logic, 200
disjunction
   Boolean, 206
   non-distributive, 240
   quantum, 240
distance
   Euclidean, *see* Euclidean
      distance, 153, 155
distance measure, 99
   and vectors, 148
document vectors, 160
double negative, 282
double-slit experiment, 219
duality, 285

edge, *see* link
Egyptian geometry, 6
Einstein, Albert, 247, *see* relativity
elements, 18, 268
   in a basis, 177
Empedocles, 21
Euclid, 7–11, *see also* axioms of
   geometry
   common notions, 8
   Elements of Geometry, 7, 25, 227
Euclidean distance, 101, 153, 155
   and cosine similarity measure,
      159
Euclidean space, 135
Eudoxus, 25
extent, 295
exterior product, 237

family tree, 263

Frege, Gottlob, 39

genera, 195
general, 69, 224
genus, 70, 76, 80
geometry
   history of, 6
   in nature, 4
   non-Euclidean, 136
Gram-Schmidt process, 230
graph, 50
   directed, *see* directed graph
   acyclic, 251
   of nouns, 53
   shortest path in, 101
graph theory, 49–53
Grassmann's Law, 235
Grassmann, Hermann, 166,
   230–237, 265, 283
Grassmannian manifold, 234
gravitation, 247
greatest lower bound, *see* meet

Hamilton, Sir William Rowan, 166
Hansard, Luke, 172
harmonic waves, 19, 219
Hasse diagram, 251
Hausdorff, Felix, 100
Hearst, Marti, 55, 195
hierarchies
   and ordered sets, 251
hierarchy or tree, 69–72, 75–79, 236
   distance in, 107
hypernym, 82
hyponym, 82, 117, 250

idioms, 61
implication, 16, 250, 257
inclination, 228
incremental clustering algorithm,
   122
Indo-European, 235
infections, 121
information retrieval
   Boolean model, 207–208
   vector model, 160–164

information theory, 97
inheritance, 72, 78
inner product, 221
integers, 34
intent, 295
intersection, 14, 206
    of intents, 275
irrelevant (in vector model), 212, *see also* orthogonal

join, 107, *see also* conjunction, *see also* disjunction, *see also* union, disjunction
    as class label, 93
    in lattice, 259

knowledge base, 81

Latent Semantic Analysis, 175–179
lattice, 247
    of numbers, 261
    of sets, 263
    of vector spaces, 264
lattice join, 259, *see* join
lattice meet, 259, *see* meet
least upper bound, *see* join
lexical acquisition, 121–123
lexicosyntactic patterns, 55, 199
Liebniz, Gottfried, 39
line, 9, 225–227
linear algebra, 135
linear span, *see* span
link, 49, 72
logic, 16
    and set theory, 16
    Aristotelean, 255–257
    Boolean, *see* Boolean logic
    mathematical, 279–283
    quantum, 237
logic gates, 205

magnitude, 105
Manhattan metric, 102, 225
manifold
    differentiable, 297
    Grassmannian, 234

mapping or map, 17
    between WORDSPACE and a taxonomy, 194
    between WORDSPACES, 185
    from WORDSPACE to graph, 199
mathematics
    definition of, 1
matrices, 145
    adjacency, 159, 198
    cooccurrence, 175
    symmetric, 159
    term-document, *see* term-document matrix
measurement, 26–31, 47, 249
    of distance, 99
    of similarity, 103
    of time, 26, 29
meet, *see also* intersection, conjunction
    in lattice, 259
meronym, 83, 115, 117
metric
    Euclidean, *see* Euclidean distance, 155
    Manhattan, *see* Manhattan metric
metric space, 100
moon, phases of, 5, 15, 27
music, 19

natural language processing, 44–49
natural numbers, 13, 33
negation, 208, *see also* vectors, negation
    Boolean, 206, 214
    double, 282
    in Aristotle, 256
network, 12, 52
Newton, Isaac, 39
Newtonian mechanics, 217
node, 49, 72
norm, of a vector, 156
normalization, 113, 150, 156–157
numbers, 18–26
    ambiguity and uniqueness, 36

object oriented database, 12

ontology, 81
operator, 136
    commutative, 137
ordered set, 248–250
    and concept lattice, 274
orthogonal, 237
    complement, 282
    subspace, 281
orthogonality, 209–214

PageRank algorithm, 74
parsing, 81
part-of-speech tags, 48
perpendicular, 106, *see also* orthogonal
phylogeny, 70
plane, 227–228, 264
Plato, 21
Platonic solids, 21–23
point, 9
points, 224
positional notation for numbers, 24
postal addresses, 77
predication, 82
probability, 203
projection, 139, 175, 184, 296
projective geometry, 233
prototype theory, 111, 130, 182
prototype-variant rule, 111
Pythagoras
    theorem of, 9, 25, 101, 153

quantifiers, 256
quantum
    Aristotelean, 224
quantum computing, 221
quantum logic, 281, 282
quantum mechanics, *see* quantum theory
quantum states, 220
quantum theory, 216–222
query, 144
    Boolean, 207
    vector, 161, 214, 230

rational numbers, 34

real numbers, 28
    as an ordered set, 248
recursion, 74
reference or zero points, 29
relation, 74
relationships, 4, 9, 42, 50, 136
    antisymmetric, 73, 75–76, 116
    symmetric, 60, 117, 160
    thesaural, 43
    transitive, 114
    and ambiguity, 118
relativity
    general theory of, 247
resistance, 104
Roget, Peter, 43
root node, 87

scalar product, 152, 221
scale, 27
Schütze, Hinrich, 172, *see also* WORDSPACE, 195, 223
semantic network, 81
semantic wormhole, 128
sense vectors, 223
set, 12–18, 74
    intersection, *see* intersection
    ordered, *see* ordered set
    subset of, 13
    union, *see* union
set theory, 224
    and Boolean logic, 207, 283
    and logic, 16
similarity measure, 103
    and vectors, 148
Singular Value Decomposition, 175, 184
smoothing, 118
span, 232, 238
species, 70, 76, 195
specific, 69, 224
Springer Link corpus, 190
square of opposition, 256, 269
state, 27
    quantum, 220
Stone Representation theorem, 207
stopwords, 173

subordinate, 265, *see also* hyponym
subspace, *see* vector subspace
subsume, 82
superordinate, 265, *see also*
    hypernym
surface, 227
syllogism, 256
symmetric relationships, *see*
    relationships, symmetric
synonymy, 42
    of measurements, 27
synset, 84

taxonomy, 72, 81, *see also* hierarchy
term-document matrix, 145, 172
thesaurus, 42–44, 53
    Roget, 43
time, 142
tokenization, 47
tokens, 47
transitive relationships, *see*
    relationships, transitive
transitivity
    and ambiguity, 118
    and Aristotle, 255
    and ordering, 251
translation and ambiguity, 193
tree, *see* hierarchy
triangle inequality, 100
troponym, 87
truth tables, 206
types, 47

UMLS, 82
Uncertainty Principle, 31, 229
union, 14, 206
    of extents, 275
unit, 27
unit circle, 105
unit vector, 157

vector space, 138
vector subspace, 264
vectors, 133–138
    addition, 134, 141
    of subspaces, 236, *see also*
        quantum disjunction
    multiplication, 137
    negation, 208–216
        and quantum states, 222
    normalized, 156, 221
    orthogonal, 209
    scalar product, 152
Venn diagram, 14
verbs, 87
vertex, *see* node

Wall Street Journal (WSJ), 168
waves
    harmonic, *see* harmonic waves
weight, 53
    of link, 56
word graph, 53
Word Spectra, 183–185
word vectors, 145
word-sense disambiguation, 215,
    220, 223
WordNet, 83–88
WORDSPACE, 168–179
    bilingual, 189
    negation in, 208
    software, 170
World Wide Web, 74
wormhole, semantic, 128

Typesetting and Programming

This work was created almost entirely using freely available software and resources.

The book was typeset by the author using Leslie Lamport's LaTeX extension of Donald Knuth's TeX typesetting system. I would like to thank Lauri Kanerva for help with the typesetting.

The word-graphs were drawn using packages from GraphViz, and the lattices were drawn using Maarten Janssen's JaLaBA. Several other graphics and photographs were kindly provided to the author free of charge and are credited in the text.

Except where otherwise credited, the graphics were drawn by the author using Xfig (these being the vast majority), Gnuplot, and Tgif.

During its development the book has been typeset on a combination of UNIX (SunOS), Linux, MacOSX and Windows machines. Screenshots were taken using several of these operating systems, and converted (printed) to PostScript and included using the epsfig package.

Many other LaTeX packages were used, including Donald Arseneau's wrapfig package which was used to generate the special interest boxes.

The models described in this book were programmed in C, Perl and Java. I would like to express enduring gratitude to Scott Cederberg, Beate Dorow, and Stefan Kaufmann, research comrades and fellow programmers. Several freely available external libraries were used, in particular Mike Berry's SVDPACK for singular value decomposition.

Other resources used in this book include WordNet (which is freely available) and several language corpora (many of which were licensed, though several corpora are now freely available on the web for research and study).

In addition to those mentioned by name, this work would not have been possible without a host of committed enthusiasts who make the fruits of their time and talent freely available to others through the open source software movement.

I will endeavour to keep a permanent collection of links to these resources and reading materials easily accessible on the web (and easily findable using the query `widdows geometry and meaning`).

Dominic Widdows
Pittsburgh, Pennsylvania
September 2004